T0261967

Natural Dyes Handbook

Natural Dyes Handbook

Edited by **Dick Baia**

New York

Published by NY Research Press,
23 West, 55th Street, Suite 816,
New York, NY 10019, USA
www.nyresearchpress.com

Natural Dyes Handbook
Edited by Dick Baia

International Standard Book Number: 978-1-63238-341-9 (Hardback)

Contents

Preface

Dyes have played a major role in the progress of the textile industry. This is a comprehensive book which aims to provide in-depth information along with a concise introduction to natural dyes. It is impossible to imagine textile materials without colorants as archaeological evidence reflects that dyeing has been extensively used for more than 5000 years. With the advancement of chemical industry, all finishing procedures of textile materials have been growing constantly and, sustainable and ecological production techniques have become extremely crucial. This book consists of results about the novel researches on natural dyeing.

The information contained in this book is the result of intensive hard work done by researchers in this field. All due efforts have been made to make this book serve as a complete guiding source for students and researchers. The topics in this book have been comprehensively explained to help readers understand the growing trends in the field.

I would like to thank the entire group of writers who made sincere efforts in this book and my family who supported me in my efforts of working on this book. I take this opportunity to thank all those who have been a guiding force throughout my life.

Editor

Part 1

Eco-Friendly Pretreatment

1

Eco-Friendly Pretreatment of Cellulosic Fabrics with Chitosan and Its Influence on Dyeing Efficiency

Mohamed Abd el-moneim Ramadan, Samar Samy,
Marwa abdulhady and Ali Ali Hebeish
Textile Research Division, National Research centre, Dokki, Giza
Egypt

1. Introduction

Wet processing of textiles uses large quantities of water, and electrical and thermal energy. Most of these processes involve the use of chemicals as assisting, accelerating or retarding their rates and are carried out at elevated temperatures to transfer mass from processing liquid medium across the surface of textile material in a reasonable time. So, we can use some natural materials and some physical tools to reduce the chemicals, water, energy and pollution. This tools and natural material can use also to improvement the fabrics surface via introduce new active groups on its. Improvement of the fabrics surface can increase the efficiency of bleaching, dyeing and finishing processes.

Chitin, a major component of the shell of crab and shrimp is one of the most abundant natural polysaccharides with a large unexplored commercial potential. Chitosan is partially or completely N-deacetylated chitin and mainly consists of B-(1,4)- linked 2- amino-2-deoxy-B-D-glucopyranose. In recent years, a number of investigations have been carried out to exploit the potential applicability of chitosan[1]. Chitosan have many applications in the medical and textile fields.

Cellulose which has been known to have good physical properties has been widely used as construction material, paper and clothes. Cellulosic fabrics can be oxidized by several oxidizing agents such as hydrogen peroxide (H_2O_2), sodium persulphate ($Na_2S_2O_8$) and potassium periodate (KIO_3). Oxidation of cellulosic fabric using sodium metaperiodate ($NaIO_4$) has been extensively investigated in the literature, since it leads to selective cleavage at the C2 and C3 vicinal hydroxyl groups to yield a product with 2,3-dialdehyde units along the polymer chain[2]. The latter is an important functional polymer for further derivatisation to specialized products.

As a naturally deriving substances, chitosan and cellulose share several common beneficial properties such as being nontoxic and biodegradable. Structurally, chitosan is slightly different from cellulose as the monosaccharide in chitosan chain is 2-amino-2-dehydroxy-D-glucose instead of D-glucose. The presence of amino group is responsible for the complete solubility of chitosan in a diluted aqueous acidic solution as a polycationic polymer whereas cellulose is totally insoluble. This polycationic nature makes chitosan very appealing as a surface treating agent for cellulose fabrics to improve its dyeability to reactive and acid dyes. Chitosan used for surface modification of cellulose fabrics. Direct padding or exhausting chitosan solution onto cotton fiber was reported to show significant improvement in dyeability of the fiber with some reactive dyes [3]. Oxidations of cotton fiber or fabric prior to the treatment of chitosan have been reported. Chitosan was applied after oxidation of cotton fabric by H_2O_2 show improved dyeability with reactive dyes [4]. Potassium periodate is known to selectivity convert 1,2-dihydroxyl groups to a pair of aldehyde groups without significant side reaction and is widely used in structural analysis of carbohydrates[2]. This oxidizing agent was used successfully for surface oxidation of cotton fiber prior to the treatment with chitosan to produce chitosan coated cotton fiber.

We undertake this work with a view to establish appropriate conditions for synthesis of cotton fabric containing chitosan. We plan to incorporate chitosan in the molecular structure of cotton surface through strong interaction between chitosan molecules and cotton molecules. To achieve the goal, creation of functional groups such as aldehyde groups is effected by $NaIO_4$ oxidation and thus obtained oxidized cotton is treated with chitosan. This is ratter a two-step process for producing cotton fabrics containing chitosan. A novel one-step process is also devised for preparation of the same modified fabrics; the fabric is treated in an aqueous solution containing the oxidant and chitosan. All modified fabrics are monitored for fixed amount of chitosan expressed as nitrogen content, carbonyl content, dyeability, strength properties and IR spectra.

2. Experimental

2.1 Materials

A bleached cotton fabric was kindly supplied by Misr Company for Spinning and Weaving, Mehallah El-Kubra, Egypt. All chemicals used in current investigation were of analytical grade. Dyes used include acid dye which was of laboratory grade and, reactive dye, namely, Procion Turquoise® MXG which was of technical grade.

2.2 Methods
2.2.1 Oxidation of cotton fabrics with sodium periodate

Unless otherwise stated, bleached cotton fabric was immersed in an aqueous solution containing different concentrations of sodium periodate (30-80 mg/100ml). The solution was stirred for 1 hr at 60 °C using a material to liquor ratio (M/L) 1:50 .The oxidized sample was washed several times with water to remove the oxidant .This oxidized sample was used for the next reaction without drying.

2.2.2 Treatment with chitosan

A chitosan solution was prepared by stirring a dispersion of chitosan (0.5-2%) in 1% (v/v) aqueous acetic acid solution. The aforementioned oxidized cotton fabric was immersed in the chitosan solution with constant shaking for different periods of time (30, 60, and 120 min.).The treatment process was performed at different temperatures (40, 60 and 80 °C). A material to liquor ratio of 1:50 was used. The treated sample was washed several times with 1% (v/v) aqueous acetic acid solution followed by water and dried at 60 °C.

2.2.3 Oxidation of cotton fabrics and chitosan with sodium periodate in one bath

An aqueous solution containing 2% chitosan, 50 mg/100ml sodium periodate was prepared. Fabric was impregnated in this aqueous solution at liquor ratio 1:50 for 1 hr at 60 °C. The cotton fabric was then washed several times with 1% (v/v) aqueous acetic acid solution followed by water and dried at 60 °C.

2.2.4 Dyeing

Dyeing with reactive dye

Cotton fabrics treated with solution periodate as described above were dyed using aqueous bath containing 1% of the reactive dye along with 5% sodium chloride and liquor ratio 1:50 at room temperature. The temperature of the dyeing bath was raised to 60 °C for 45 minutes. The fabrics were then rinsed with water and treated with an aqueous solution containing 1% wetting agent at 60 °C for 30 minutes at a liquor ratio 1:50. The dyed fabrics were rinsed with hot water followed by cold water and finally dried at ambient conditions

Dyeing with acid dye

After being treated with solution periodate, the cotton fabrics samples were dyed in an aqueous bath containing 1% of the acid dye together with 2% sodium sulphate using a liquor ratio 1:50 at pH, 5-6 and temperature 40 °C. The latter was then raised to 100 °C for 60 minutes. At this end, the fabrics were squeezed, washed with water and treated with an aqueous solution containing 2% wetting agent at 60 °C for 30 minutes. The dyed fabrics so obtained were rinsed with hot water followed by cold water and finally dried at ambient conditions.

2.2.5 Testing and analysis

- The Nitrogen content was determined according to Kjeldahl method[5]
- The carbonyl content, expressed as meq./100g fabric, was determined according to a reported method [6].
- Tensile strength and elongation at break were measured using the strip method according to ASTM [7]
- The colour strength of dyed fabrics, expressed as K/S, was determined as described elsewhere [8].
- FT-IR spectroscopy

The FT-IR spectra of cotton fabrics treated with $NaIO_4$ and chitosan in two subsequent steps as well as concurrently were recorded on a Nexus 670 FT-IR spectrophotometer, Nicolet, USA, in the spectra range 4000-400 cm^{-1} using the KBr disc technique.

3. Results and discussion

Complexation of chitosan with cotton cellulose to produce cotton –based new textiles is of paramount concern. To render these new cotton textiles more durable and stable for subsequent treatments such as dyeing and finishing, it is a must to induce strong interactions between chitosan and cotton. One of the approaches to achieve this is to create aldehyde groups in the molecular structure of cotton cellulose. These aldehyde groups undergo coupling with the amino groups of chitosan to form iminic bonds [1] whereby chitosan is fixed to cotton surface through a series of reactions as shown under:

Our work involves fixation of chitosan to bleached cotton fabric as per two methods. The fist method is based on oxidation of the fabric with $NaIO_4$ in acidic medium to yield fabric containing 2,3-vicinal diol of the glucose unit of cotton cellulose or what is called dialdehyde cellulose. The so obtained oxidized fabric is then subjected to chitosan treatment where interaction occurs thereby causing fixation of chitosan on the fabric surface. In the second method, the fabric is treated in a bath containing $NaIO_4$ and chitosan where simultaneous oxidation and chitosan fixation take place.

When dealing with the second method, mention should be made of the effect of $NaIO_4$ on the chitosan. According to previous reports[9], the chitosan molecule is susceptible to oxidation by sodium periodate at five points: the terminal aldehyde group at C1, the secondary hydroxyl on C4 at the non reducing end of the chain, the secondary hydroxyl on C3, the primary hydroxyl at C6 and the amino group at C2 position. In addition molecular chain scission of chitosan occurs through the attack of the oxidant on the 1-4 glucosidic linkages. When chitosan is oxidized the action takes place chiefly at C3 and C6 hydroxyls, the amino group at C2 and at the 1-4 glucosidic linkage by virtue of the greater abundance of these sites as compared with the two types of terminal groups.

With the above in mind, factors affecting the major technical properties of cotton-based new textile brought about by the first and second methods were thoroughly investigated. Factors studied include concentrations of chitosan and $NaIO_4$ as well as time and temperature of the treatments. On the other hand, the obtained cotton products were monitored for nitrogen content, carbonyl content, tensile strength, elongation at break and dyeability. IR spectra for cotton fabrics-containing chitosan which were processed as per the two methods are also presented.

3.1 Sodium periodate concentration

Table 1 shows the effect of sodium periodate concentration on the amount of fixed chitosan (expressed as nitrogen content), carboxyl content, tensile strength and elongation at break of the oxidized fabric. Obviously the amount of fixed chitosan on the cotton fabric increases by increasing the oxidant concentration up to 50 mg/100ml. Further increase in the oxidant concentration to 80 mg/100ml causes no significant increase in nitrogen content. This phenomena could be explained by considering the difference in the reaction site of the oxidation and Schiff's base formation in the cellulosic fabric. During the oxidation, the small periodate ion might be able to enter the cellulosic fabric interior and the glucose unit both inside and on the surface of the cellulosic fabric may be oxidized. On the other hand, chitosan is a huge molecule that cannot enter the fabric, and the modification with chitosan occurs on the surface of the fabric. Results of table 1 reveal that the carbonyl content of the fabric in question, i.e. periodate-oxidized fabric containing chitosan, increases by increasing the oxidant concentration. This is rather a manifestation of oxidation of hydroxyl groups of the cotton fabric to aldehydic groups under the progressive action of the oxidant at high concentrations.

Tensile strength decreases substantially after the fabric was subjected to periodate treatment at a concentration of 30 mg/100ml followed by chitosan treatment. Increasing the periodate concentration up to 80 mg/100ml causes no further significant decrease in the tensile strength of periodate oxidized fabric-containing chitosan. While the loss in tensile strength of the fabric upon using sodium periodate at 30 mg/100ml concentration could be interpreted in terms of degradation of cotton cellulose, the no significant change in tensile strength upon using higher concentrations calls for extra strength imported to the fabric by higher amounts of fixed chitosan as evidenced by the higher contents. The chitosan molecules seems to form a film fixed on the cotton cellulose (in the fabric form) through strong interactions thereby compensating for the higher losses in tensile strength expected at higher sodium periodate concentrations. It is also possible that the cotton cellulose undergoes modification during the initial oxidation and such modification makes the cellulose less susceptible for farther oxidation particular via chain scission. On the other hand, elongation at break marginally reduced after oxidation and chitosan treatment regardless of the oxidant concentration used within the range studied.

$[NaIO_4]$ mg/100ml	Nitrogen content (%)	Carbonyl content Meq/100g S	Tensile strength (Newton)	Elongation at Break (%)
0.0	0.112	15.122	38.25	11.34
30	0.181	20.098	30.17	11.02
50	0.243	25.105	29.53	10.71
80	0.252	31.167	29.42	10.45

Conditions used for oxidation of cotton fabric: time,1hr; temperature,60 ᵒC, M/L, 1:50
Condition used for chitosan treatment: [chitosan], 1%; time, 1.5 hr; temp, 60ᵒC; M/L, 1:50.

Table 1. Effect of sodium periodate concentration on some chemical and mechanical properties of chitosan-containing cotton fabric.

The above findings indicate that modified cotton fabrics which enjoy the presence of chitosan as evidenced by their nitrogen content and acidic properties as evidenced by the carbonyl content while retaining much of their strength properties can be achieved using $NaIO_4$ concentration 30-5- mg/100ml followed by treatment with chitosan at a concentration of 1%.

3.2 Chitosan concentration

Table 2 shows the dependence of the modification effect- expressed as nitrogen content, carboxyl content and strength properties including tensile strength and elongation at break on the chitosan concentration. As is evident the nitrogen content and strength properties of modified fabric increase as the chitosan concentration increases. At higher concentrations, chitosan molecules would be greatly available in the proximity of the macrostructure of cotton and fabric surfaces both containing aldehyde groups thereby leading to Schiff's base formation between aldehyde groups and chitosan amino groups.

[chitosan] %	Nitrogen content (%)	Carbonyl content Meq/100g S	Tensile strength (Newton)	Elongation at Break (%)
0.0	0.000	36.231	28.22	10.31
0.5	0.191	32.143	28.52	11.02
1.0	0.243	25.105	29.53	11.71
1.5	0.292	22.125	31.21	12.44
2.0	0.324	20.176	31.51	12.53

Conditions used for cotton fabric oxidation: [$NaIO_4$], 50 mg/100ml; time, 1hr; temp, 60 ᵒC, M/L, 1:50
Condition used in chitosan treatment: time, 1.5 hr; temp, 60 ᵒC; M/L, 1:50

Table 2. Dependence of the magnitude of modification of cotton fabric on chitosan concentration.

It is further observed that the values of the carbonyl content of the modified fabric decrease substantially by increasing chitosan concentration. This indicates that the carboxyl groups of the modified cotton fabric under investigation are involved in chemical reactions with chitosan. Once this is the case, the carbonyl groups are masked and their values decreases by increasing the chitosan concentration where opportunities of interactions are better.

3.3 Temperature of chitosan treatment

Table 3 discloses the effect of chitosan treatment temperature of oxidized cotton on some chemical and mechanical properties of the modified fabric. The treatment was carried out at different temperatures for 90 minutes using chitosan concentration of 2%. The results signify that the nitrogen content increases by raising the chitosan treatment temperature from 40°C to 80 °C .This could be ascribed to increased amount of incorporated chitosan into the oxidized fabric as a result the favorable effect of temperature on swallability of cotton and mobility of chitosan molecules; both enhance the magnitude of interactions between cotton and chitosan. On the other hand, raising the chitosan treatment temperature adversely affects the tensile strength of the modified fabric. Most probably greater penetration of the highly mobile chitosan at higher temperatures in the fibriller structure of swollen cotton causes rigidity and, in term, decrement in tensile strength. The results of elongation at break are in confirmation with this.

Temp. °C	Nitrogen content (%)	Carbonyl content Meq/100g S	Tensile strength (Newton)	Elongation at Break (%)
40	0.196	24.116	34.21	13.13
60	0.324	20.176	31.51	12.53
80	0.513	18.523	29.53	11.71

Conditions used for oxidation of cotton fabric: [NaIO₄], 50 mg/100ml; time, 1hr; temp, 60 ºC, M/L, 1:50
Condition used in chitosan treatment: [chitosan], 2%; time, 1.5 hr; M/L, 1:50

Table 3. Effect of chitosan treatment temperature of the oxidized cotton on major technical properties of the modified fabric

Table 3 depicts that the carbonyl content decreases by raising the chitosan treatment temperature from 40°C to 80 °C. This state of affairs implies that higher temperature acts in favour of the interactions of chitosan with cotton cellulose and that these interactions involve, inter alia, the carbonyl groups of cotton.

3.4 Time of chitosan treatment

Table 4 shows the effect of time of chitosan treatment on major technical properties of the obtained modified cotton fabrics. It is seen that the nitrogen content increases by prolonging the time of chitosan treatment within the range studied. Provision of better contact and intimate association of chitosan with cotton cellulose at longer duration period would account for this. On the other hand, results of carboxyl content, feature that the carbonyl content decreases by increasing the time of chitosan treatment. This is rather similar to the effect of chitosan treatment temperature discussed above and could be explained on the same basis.

Time (hr)	Nitrogen content (%)	Carbonyl content Meq/100g S	Tensile strength (Newton)	Elongation at break(%)
0.5	0.237	23.335	31.32	13.14
1.0	0.272	19.812	30.52	12.53
1.5	0.513	18.523	29.53	11.71

Conditions used for oxidation of cotton fabric: [NaIO₄], 50 mg/100ml; time, 1hr; temp, 60 °C, M/L, 1:50
Condition used in chitosan treatment: [chitosan], 2%; temp, 80 °C ; M/L 1:50

Table 4. Effect of time of chitosan treatment of oxidized fabric on some chemical and mechanical properties of the obtained modified fabric

Table 4 shows that the tensile strength decreases from ca 31 to ca 29 Newton when the duration of chitosan treatment increases from o.5 to 1.5 hr. This little effect of time of chitosan treatment is also seen with respect to elongation at break. At any event, however, these decrements are considered to be a direct consequence of rigidity conferred on cotton caused by chitosan penetration. As already stated, rigidity is proportionally related to the amount of chitosan on the fabric and the extent of penetration of former in the latter.

3.5 One step method for modification

In the foregoing sections, innovative modified cotton textiles could be achieved by a two-step method namely, periodate oxidation of cotton fabrics in one step followed by treatment of these fabrics in a second step by chitosan. In order to avoid detrimental effects of oxidation and in order to save time, chemicals and energy the two steps were combined in a single stage process where the cotton fabrics were treated in an aqueous solution containing the periodate oxidant and the chitosan under conditions emanated from the studies of the factors discussed above.

The nitrogen content, the carbonyl content and strength properties of cotton fabrics before and after being processed as per the two-step process and the one-step process are set out in table 5. For convenience the untreated cotton fabric and the two-step processed modified cotton fabric and the one-step processed cotton fabrics will be referred to as substrate I, substrate II and substrate III, respectively.

Substrate	Nitrogen content (%)	Carbonyl content Meq/100g S	Tensile strength (Newton)	Elongation at Break (%)
I	0.000	17.341	38.32	11.3
II	0.513	18.523	29.53	11.71
III	0.652	16.535	45.12	13.52

Where Substrate I: bleached cotton fabric; substrate II: two-step processed modified cotton; substrate III: one –step processed modified cotton. Two-step process involves oxidation by NaIO₄ and treatment with chitosan in two consecutive steps. One-step process involves treatment of the cotton fabric with an aqueous solution containing oxidant and chitosan.
Conditions used for oxidation of cotton fabric: [NaIO₄], 50 mg/100ml ;time,1hr; temp,60 °C, M/L, 1:50
Condition used for chitosan treatment: [chitosan], 2%; temp, 80 °C ; M/L 1:50
Condition used in one -step process: [NaIO₄], 50 mg/100ml; [chitosan], 2%; temp, 80 °C ; M/L 1:50

Table 5. Comparison among substrates I, II, III.

A comparison among the substrates I , II and III with respect to nitrogen content, tensile strength and elongation at break and carbonyl content as shown in table 5 would reveal that substrate II exhibits the carbonyl content and the lowest tensile strength as compared with substrates I and III; indicating the determinal effect of oxidation prior to chitosan treatment. On the contrary, substrate III acquires the highest nitrogen content and the highest tensile strength and elongation at break while retaining a carboxyl content which is equal to that of substrate I. This, indeed, signifies the advantages of the one-step process in avoiding prior oxidation entailed in the two-step process.

Presence of chitosan during oxidation of the cotton fabric with $NaIO_4$ seems to protect the cotton against oxidation. At the same time $NaIO_4$ oxidizes chitosan to produce chitosan products with better solubility and more uniform structure and, in turn, better film-forming properties. Such properties will be reflected on strength of the film and the extra strength brought about thereof when the cotton fabric is coated with this film. Achieving substrate III with its potential properties is regarded to be the most salient output of the current work.

Dyeability

Substrate I, substrate II and substrate III were dyed independently with reactive and acid dyes. The results obtained are shown in table 6. It is observed that regardless of the dye used. Substrate III exhibits colour strength which is higher than those of substrate II. This again signifies the superiority of substrate III which, indeed, together with its other properties (table 5) advocate the one-step process for preparation of innovative modified cotton fabrics-containing chitosan.

Dye	Colour strength (K/S)		
	Substrate 1	Substrate II	Substrate III
Reactive dye	16.4	22.78	23.05
Acid dye	2.8	4.47	4.49

Substrate 1: Bleached cotton fabric
Substrate II: processed as per the two-step process
Substrate III: processed as per the one-step process

Table 6. Dyeing of the different substrates under investigation with reactive and acid dyes

IR analysis

The IR spectrum of substrate II, as shown in figure 1, discloses the presence of a broad band at 3300-3500 cm^{-1} attributable to the NH_2 and OH groups , a weak absorption band appeared at nearly 1730 cm^{-1} due to the stretching vibration of the C=O double bond of the aldehydic group and a strong absorption band at 1641 cm^{-1} which is assigned to the C=N group was formed between the aldehydic group and chitosan. Also the IR spectrum of substrate III, as shown in figure 2 reveal the presence of a broad band at 3300-3500 cm^{-1} attributable to the NH_2 and OH groups , a strong absorption band at 2644 cm^{-1} which is assigned to the C=N group and a weak absorption band at nearly 1730 cm^{-1} foe C=O group.. These IR spectrums confirm the presence of chitosan in both substrates. That is, chitosan interacted with preoxidized cotton fabric as well as with cotton fabric subjected to concurrent oxidation and chitosan treatments.

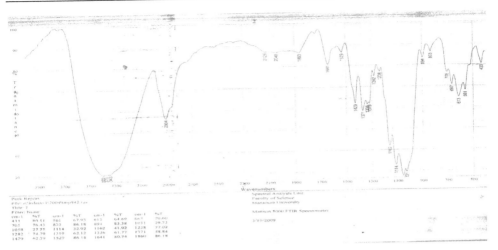

Fig. 1. IR spectrum of substrate II

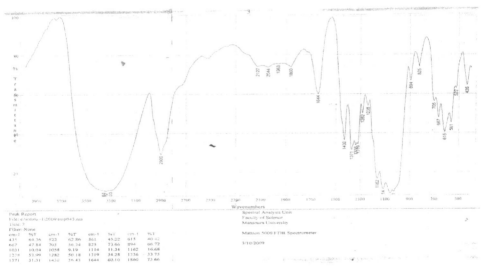

Fig. 2. IR spectrum of substrate III

4. Conclusion

Chemical modification of cotton cellulose in the fabric form was effected through periodate oxidation treatment and chitosan treatment in the consecutive steps under different conditions. A single-step process was also devised for preparation of the same modified cotton. The idea in both cases was to create functional groups in the molecular structure of cotton such as aldehyde and carboxyl groups to expedite strong interactions with chitosan. Modified cotton fabrics processed as per the two processes were monitored, nitrogen content, carbonyl content, tensile strength and elongation at break in addition IR spectra.

Results conclude that modified cotton fabrics processed as per the one-step process is by for better than those processed according to the two-step process. Attachment of chitosan to the fabric considerably improves the dye uptake of reactive and acid dyes resulting in greater the colour yield (K/S) as compared with the nonmodified fabric. The enhanced dyeability of the modified fabrics is attributed the reduction of the culombic repulsion between the fabric surface and the anionic dye molecules in the presence of the positively charged chitosan on the surface. The drop in fabric strength by oxidation by oxidation prior to chitosan treatment could be overcome by combining the oxidation and chitosan treatment in one-step process where the fabric is treated in an aqueous solution containing $NaIO_4$ and chitosan.

5. References

[1] Liu, X.D., Nishi, N., Tokura, S., & Sakairi, N. (2001). Chitosan coated cotton fiber: preparation and physical properties. Carbohydrate Polymers, 44, 233-238.
[2] Varma, A.J. & Kulkarni, M.P. (2002). Oxidation of cellulose under conrolled conditions. Polymer Degradation and Stability, 77, 25-27.
[3] Shin,Y., & Yoo, D.I. (1998). Use of chitosan to improve dyeability of DP-finished cotton (II). Journal Applied Polymer Science, 67, 1515-1521.
[4] Weltrowski, M., & Masri, M.S. (1996). Method for treatment of cellulosefabrics to improve their dyeability with reactive dyes. US Patent 5, 501, 711, 26 March.
[5] D. Duchene and D. Wouessidjewe, Acta pharm. Technol., 36, 1 (1991)
[6] Klimova, V. A. & Zabrodina, K. S. (1966), Zhakn, 15, 726.
[7] ASTM Test Methods D-1628, in "Book of Standard", ASTM, Philadelphia, PA, Part 24, 1972.
[8] Kubellka, P., Munk, F. & Tech, Z. (1931). Physik.12,593.
[9] Hebeish, A.A., Waly, A., Higazy, A. & Elshafei A. (2004). Hydrolytic and Oxidative Degradation of Chitosan. Egyt.J.Chem.Special Issue, 102-122.

Part 2

Dyeing

2

Dyeing of Textiles with Natural Dyes

Ashis Kumar Samanta and Adwaita Konar
Department of Jute and Fibre Technology,
Institute of Jute Technology, University of Calcutta
India

1. Introduction

Textile materials (natural and synthetic) used to be coloured for value addition, look and desire of the customers. Anciently, this purpose of colouring textile was initiated using colours of natural source, untill synthetic colours/dyes were invented and commercialized. For ready availability of pure synthetic dyes of different types/classes and its cost advantages, most of textile dyers/ manufacturers shifted towards use of synthetic colourant. Almost all the synthetic colourants being synthesized from petrochemical sources through hazardous chemical processes poses threat towards its eco-friendliness.

Hence, worldwide, growing consciousness about organic value of eco-friendly products has generated renewed interest of consumers towards use of textiles (preferably natural fibre product) dyed with eco-friendly natural dyes. Natural dyes are known for their use in colouring of food substrate, leather as well as natural fibres like wool, silk and cotton as major areas of application since pre-historic times. Although this ancient art of dyeing textiles with natural dyes withstood the ravages of time, but due to the wide availability of synthetic dyes at an economical price, a rapid decline in natural dyeing continued. However, even after a century, the uses of natural dyes never erode completely and they are being still used in different places of the world. Thus, natural dyeing of different textiles and leathers has been continued mainly in the decentralized sector for specialty products besides the use of synthetic dyes in the large scale sector for general textiles/apparels.

Recently, most of the commercial dyers and textile export houses have started re-looking to the maximum possibilities of using natural dyes for dyeing and printing of different textiles for targeting niche market. Natural dyes produce very uncommon, soothing and soft shades as compared to synthetic dyes. On the other hand, synthetic dyes, which are widely available at an economical price and produce a wide variety of colours, sometimes causes skin allergy and other harmfulness to human body, produces toxicity/chemical hazards during its synthesis, releases undesirable/hazardous/toxic chemicals etc.

For successful commercial use of natural dyes for any particular fibres, the appropriate and standardized techniques for dyeing for that particular fibre-natural dye system need to be adopted. Therefore to obtain newer shade with acceptable colour fastness behaviour and reproducible colour yield, appropriate scientific dyeing techniques/procedures are to be derived. Thus, relevant scientific studies and its output on standardization of dyeing methods, dyeing process variables, dyeing kinetics and test of compatibility of selective natural dyes have become very important, however the information on which is insufficient. That is why, this chapter is very much relevant to the current need of the textile dyers. An attempt has been made here to give scientific overview on dyeing of textiles with natural dyes and related issues.

2. Definition of natural dyes/colouants

The word 'natural dye' covers all the dyes derived from the natural sources like plants, animal and minerals. Natural dyes are mostly non-substantive and must be applied on textiles by the help of mordants, usually a metallic salt, having an affinity for both the colouring matter and the fibre. Transition metal ions usually have strong co-ordinating power and/or capable of forming week to medium attraction/interaction forces and thus can act as bridging material to create substantivity of natural dyes/colourants when a textile material being impregnated with such metallic salt (i.e. mordanted) is subjected to dyeing with different natural dyes, usually having some mordantable groups facilitating fixation of such dye/colourant. These metallic mordants after combining with dye in the fibre, it forms an insoluble precipitate or lake and thus both the dye and mordant get fixed to become wash fast to a reasonable level.

3. Advantages and disadvantages of natural dyes/ colouants

In the recent years, there has been a trend to revive the art of natural dyeing. This is mainly because in some aspects natural colourants are advantageous against synthetic dyes. Some of these advantages along with some limitations (disadvantages) are listed below:

3.1 Advantages of natural dyes/ colouants

i. The shades produced by natural dyes/colourants are usually soft, lustrous and soothing to the human eye.
ii. Natural dyestuff can produce a wide range of colours by mix and match system. A small variation in the dyeing technique or the use of different mordants with the same dye (polygenetic type natural dye) can shift the colours to a wide range or create totally new colours, which are not easily possible with synthetic dyestuffs.
iii. Natural dyestuffs produce rare colour ideas and are automatically harmonizing.
iv. Unlike non-renewable basic raw materials for synthetic dyes, the natural dyes are usually renewable, being agro-renewable/vegetable based and at the same time biodegradable.
v. In some cases like harda, indigo *etc.*, the waste in the process becomes an ideal fertilizer for use in agricultural fields. Therefore, no disposal problem of this natural waste.
vi. Many plants thrive on wastelands. Thus, wasteland utilization is an added merit of the natural dyes. Dyes like madder grow as host in tea gardens. So there is no additional cost or effort required to grow it.
vii. This is a labour intensive industry, thereby providing job opportunities for all those engaged in cultivation, extraction and application of these dyes on textile/food/leather etc.
viii. Application of natural dyes has potential to earn carbon credit by reducing consumption of fossil fuel (petroleum) based synthetic dyes.

ix. Some of its constituents are anti-allergens, hence prove safe for skin contact and are mostly non-hazardous to human health.
x. Some of the natural dyes are enhanced with age, while synthetic dyes fade with time.
xi. Natural dyes bleed but do not stain other fabrics, turmeric being an exception.
xii. Natural dyes are usually moth proof and can replace synthetic dyes in kids garments and food-stuffs for safety.

Despite these advantages, natural dyes do carry some inherent disadvantages, which are responsible for the decline of this ancient art of dyeing textiles.

3.2 Limitation/ disadvantages of natural dyes/ colouants

i. It is difficult to reproduce shades by using natural dyes/colourants, as these agro-products vary from one crop season to another crop season, place to place and species to species, maturity period etc.
ii. It is difficult to standardize a recipe for the use of natural dyes, as the natural dyeing process and its colour development depends not only on colour component but also on materials.
iii. Natural dyeing requires skilled workmanship and is therefore expensive. Low colour yield of source natural dyes thus necessitates the use of more dyestuffs, larger dyeing time and excess cost for mordants and mordanting.
iv. Scientific backup of a large part of the science involved in natural dyeing is still need to be explored.
v. Lack of availability of precise technical knowledge on extraction and dyeing techniques.
vi. The dyed textile may change colour when exposed to the sun, sweat and air.
vii. Nearly all-natural dyes with a few exceptions require the use of mordants to fix them on to the textile substrate. While dyeing, a substantial portion of the mordant remains unexhausted in the residual dye bath and may pose serious effluent disposal problem.
viii. With a few exceptions, most of the natural dyes are fugitive even when applied in conjunction with a mordant. Therefore, sometimes their colour fastness performance ratings are inadequate for modern textile usage.

4. Classification of natural dyes/ colouants

Natural dyes can be classified (Gulrajani & Gupta, 1992) in a number of ways. The earliest classification was according to alphabetical order or according to the botanical names. Later, it was classified in various ways, e.g. on the basis of hue, chemical constitution, application class etc.

a. In "treatise on permanent colours" by Bancroft, natural dyes are classified into two groups: 'Substantive Dyes' such as indigo, turmeric etc. which dye the fibers directly and 'Adjective Dyes' such as logwood, madder etc. which are mordanted with a metallic salt.
b. Humme classify the colouring matter as 'Monogenetic Dyes', those produce only one colour irrespective of the mordant present on the fibre or applied along with the dye and 'Polygenetic Dyes', those produce different colour with different mordant applied, e.g., alizarin (Dedhia, 1998)
c. In the colour index the natural dyes are classified according to the hue (Predominating colour). The number of dyes in each hue are as follows in table-1:

CI Natural	No. of Dyes	Percent
Yellow	28	30.4
Orange	6	6.5
Red	32	34.8
Blue	3	3.3
Green	5	5.5
Brown	12	13
Black	6	6.5

Table 1. Showing the number of natural dyes in each hue as per the colour index.

On the basis of hues, natural dyes can be classified as follows:

i. Red colour dyes: most red dyes are hidden in roots or barks of plants or camouflaged in the bodies of dull grey insects. They are almost invariably based on anthraquinone and its derivatives. These dyes are stable to light and washing.

ii. Yellow colour dyes: Yellow is the liveliest and perhaps the most abundant of all hues in nature. About 90% of the yellow dyes are flavonoids. Generally, they produce pale shade with quicker fading except turmeric, which produce dull deep shade but considered to be susceptible to light as they emit fluorescence. Wash fastness rating of natural yellow dyes ranges from fair to excellent, e.g., tesu, turmeric, kapila.

iii. Blue colour dyes are indigo and woad, give excellent fastness to light and washing.

iv. Black colour dyes: Black shades, generally obtained from tannin rich plant natural dyes and appreciably substantive towards cellulosic and protein fibre, imparts good overall fastness properties. Examples – logwood, harda, custard apple etc.

d. On the basis of origin, natural dyes are broadly classified into three categories: vegetable, mineral and animal origin. About 500 vegetable origin dyes, colouring matter derived from root, leaf, bark, trunk or fruit of plants, are as follows in table-2

Part of the Plants	Dyestuffs
Root	Turmeric, Madder (Manjistha), Onions, Beet-root
Bark/ Branches	Purple bark, Sappan wood, Shillicorai, Khair, Red, Sandalwood
Leaf	Indigo, Henna, Eucalyptus, Tea, Cardamon, Coral Jasmine, Lemon Grass
Flowers (Petals)	Marigold, Dahlia, Tesu, Kusum
Fruits/Seeds	Latkan, Pomegranate rind, Beetle nut, Myrobolan (Harda)

Table 2. Showing some common natural dyestuffs obtained from different vegetable origin.

Mineral origin colourants are derived from specific mineral natural source or so-called mineral colours are produced from purified inorganic compounds. Some of the important mineral colourants are chrome-yellow, iron-buff, narkin-yellow, Prussian-blue and manganese brown. Animal origin lac, cochineal and kermes have been the principal natural dyes yielding from the insects.

e. Natural dyes can also be classified on the basis of their chemical constitution (Dedhia, 1998).

  i. Indigoid dyes: Indigo and tyrian purple are the most common examples of this class. Another blue dye, woad also possesses indigo as the main dyeing component.

  ii. Anthraquinone dyes: Almost all the red natural dyes are based on the anthraquinoid structure having both plant and mineral origin. Madder, lacs, kermes, cochineal are some of the dyes possess this type of structure. These are generally mordant dyes.

  iii. Alphanaphthoquinones: Typical example of this class is lawsone (henna), cultivated mainly in India and Egypt. Another similar dye is juglone, obtained from the shells of unripe walnuts. These dyes are generally disperse dyes and give shades of orange.

  iv. Flavonoids, which yield yellow dyes can be classified under flavones, isoflavones, aurones and chalcones. Flavones are colourless organic compounds. Most of the natural yellows are derivatives of hydroxyl and methoxy substituted flavones and isoflavones. Common example is weld (containing luteolin pigment) giving brilliant and fast colours on both wool and silk.

  v. Di-hydropyrans: Closely related in chemical structure to the flavones are substituted di-hydropyrans, *viz.* haematin and its leuco form, haematoxylin. These are important natural dyes for dark shades on silk, wool and cotton. Logwood, brazil wood and sappan-wood are the common example.

  vi. Anthocyanidins: The naturally occurring member of this class includes carajurin, a direct orange dye for wool and cotton. It is obtained from the leaves of bignonia chica.

  vii. Carotenoids: The class name carotene is derived from the orange pigment found in carrots. In these, the colour is due to the presence of long conjugated double bonds.

f. Another method of classifying natural dye is on the basis of the method of application (Gulrajani & Gupta, 1992).

  i. **Mordant dyes** are dyestuffs which require a mordant in their application as they have no affinity for the fiber being dyed. A mordant dye should have electron donating groups capable of forming a complex with the transition metal salt, e.g., madder, fustic, persian, berries, kermes, cochineal etc.

  ii. **Vat dyes** are water insoluble dyes which are first converted to their water soluble form (reducing with Na-hydrosulphite and then solubilising it with alkali) and then applied to the fibres. The true colour is produced only on oxidation followed by treatment with a hot soap solution, *e.g.*, indigo.

  iii. **Direct dyes** are those dyes that have tremendous affinity for the cellulosic fibres. They are dyed from a boiling dye bath. Turmeric, harda, pomegranate rind etc. are the few of the direct natural dyes.

  iv. **Acid dyes** are applied from an acidic medium. The dye molecules have either sulphonic or carboxylic group (s) which can form an electrovalent bond with amino groups of wool and silk. An after treatment with tannic acid known as back tanning improves the fastness of these type of dyes, *e.g.*, saffron.

  v. **Disperse dye** has a relatively low molecular mass, low solubility and no strong solubilizing groups. Disperse dyes can be applied on to hydrophobic synthetic fibre from neutral to mildly acidic pH. They can also be applied to silk and wool. These dyes can be post-mordanted with chromium, copper and tin salts, *e.g.*, lawsone and many other flavone and anthroquinone dyes.

vi. **Basic or cationic dyes** on ionization give coloured cations and form an electrovalent bond with the –COOH group of wool and silk. These dyes are applied from neutral to mildly acidic pH. These dyes have poor light fastness, *e.g.*, berberine.

5. Extraction process of colour component from natural dyes

Extraction of colour component from source natural dye material is important step for dyeing any textile substrate to maximize the colour yield. Moreover, standardization of extraction process and optimizing the extraction variables both, for a particular source natural dye material have technical and commercial importance on colour yield and cost of extraction process as well as dyeing cost. The natural dyes can be taken from various vegetable sources like flowers, stem or wood, roots, bark, etc. as well as animal sources and mineral sources. The colour component present in these sources needs to be extracted so that it can be applied suitably on textiles. Natural dyes of different origin can be extracted using aqueous method i.e. by using water for the extraction with or without addition of salt/acid/alkali/alcohol in the extraction bath, supercritical fluid extraction, enzyme assisted extraction, alcoholic/organic solvent extraction by using relevant extracting equipment or soxhlet extraction method with use of alcohol and benzene mixture and finally to filterate, evaporate and to dry using ultra filtration equipment or centrifuge rotatory vacuum pump/or by extraction under reduced pressure. Now a days, there has been industrial methods available for extracting colour components/purified colour substances from natural dyes for their easy applications.

The collected source material is generally shadow dried in air or sun dried within a temperature range of 37-40°C for the moisture content of the source natural dye material is reduced to 10-15% with proper drying since most of the material have moisture content of 40-80% and can not be stored without drying. After drying, grinding is carried out to break down the material into very small units or preferably powder form. Extraction refers to separating the desired colour component by physical or chemical means with the aid of a solvent. Optimum conditions of extraction variables are determined through extracting the natural colour component from source material by varying extraction parameters of liquor and measuring the optical density of corresponding coloured liquor by using spectrophotometer. Also, the gravimetric yield of colour can be measured by filtering the extraction liquor through standard filtration process followed by evaporation of solvent, washing and finally drying to get the purified natural colour.

5.1 Aqueous extraction system

For optimizing the extraction method of colour component in aqueous medium, dried and finely cut source material of natural dye is grinded in powdered form and then the colour component is extracted in water employing a standard process. The aqueous extraction of dye liquor is carried out under varying condition, such as time of extraction, temperature of extraction bath, pH of extraction liquor, concentration of colour-source material (powdered form of source natural dye material) and Material-to-liquor ratio (MLR). In each case, the optical density or absorbance value at a particular (maximum) absorbance wavelength for the aqueous extract of the natural dye material can be estimated using UV-Vis absorbance spectrophotometer.

Many scientists have reported the optimized process of extraction of natural dyes from source. Colour from leaves of eucalyptus hybrid, seeds of cassia tora and grewia optiva are extracted by using aqueous medium under varying conditions (Dayal & Dobhal, 2001). Natural dyes are extracted from biomass products namely cutch, ratanjot, madder (Khan et al, 2006) and from hinjal, jujube bark (Maulik & Pradhan, 2005) in aqueous medium. An attempt has been made to extract natural dye from the coffee-seed for its application in dyeing textiles like cotton and silk (Teli & Paul, 2006). Grey jute fabric is dyed with extracts from deodar leaf (Pan et al, 2003) jackfruit wood and eucalyptus leaf by soaking it soft water and boiling it for 4 hours separately. Extraction (Verma & Gupta, 1995) of natural dyes is also reported from overnight soaked wattle bark in distilled water followed by boiled it in pressure vessel and filtered it to obtain a residual dye powder of about 15 to 20 % (w/w) of the bark. Colours are extracted from marie gold and chrysanthemum flowers by boiling the dry petals with acidified or salt water and reported it to be the best (Deo & Paul, 2000; Sarkar et al 2005 & 2006; Saxena et al, 2001). Natural colour extraction process has also been optimized in aqueous media for various source natural dye materials as follows (Konar, 2011):

- Pomegranate Rind: Pre-cut and dried rind is initially crushed to powder form and then it is extracted in water using an optimized condition of extractions using MLR- 1:20, temperature -90°C and time - 45 min and then it is filtered to obtain approximately 40% (w/w) clear extract of coloured aqueous solution of pomegranate rind having pH 11.

- Mariegold (Genda): Dried petal of mariegold is initially crushed to powder form and then extracted in water using an optimized condition of extraction using MLR 1:20 at 80°C for 45 min at pH 11 and then it is filtered to obtain approximately 40% (w/w) coloured aqueous extract of mariegold.

- Babool (Babla): Sun- dried chips (pre-cut) of babool bark is initially crushed to powder form and then it is extracted in water using an optimized condition of extractions, by boiling in water at 100°C for 120min. and using MLR 1:20 and then it is filtered to obtain 40% (w/w) clear extract of coloured aqueous solution of babool having pH 11.

- Catechu (Khayer): Pre-dried powder of catechu is initially crushed to powder form and then extracted in aqueous medium using an optimized condition of extractions by heating in water bath at 90°C having MLR 1:20 and then it is filtered to obtain 40% (w/w) extract of coloured aqueous solution of catechu having pH 12.

- Jack fruit wood: Pre-cut and dried chips of jack fruit wood is initially crushed to powder form and then colour is extracted in water using an optimized conditions of extractions by boiling in water at 100°C for 30 minutes and using ML ratio (MLR) 1:10 and then it is filtered to obtain 40% (w/w) clear extract of coloured aqueous solution of jack fruit wood having pH 11.

- Red sandal wood: Dried pre-cut chips are crushed to powder form and colour is extracted under optimized conditions by heating it in water at 80°C for 90 minutes at pH 4.5 and MLR 1:20.

5.2 Extraction by non-aqueous and other solvent assisted system

Due to increasingly stringent environmental regulations, supercritical fluid extraction (SFE) has gained wide acceptance in recent years as an alternative to conventional solvent extraction for separation of organic compound in many analytical and industrial process. In recent past decade, SFE has been applied successfully to the extraction of a variety of organic compounds from herbs, other plant material as well as natural colourant from source natural dye material. With increasing public interest in natural products, SFE may become a standard extraction technique for source natural dye material and other herbs and food items. Supercritical fluid extraction using carbon dioxide as a solvent has provided an excellent alternative to the use of chemical solvents. Over the past three decades, supercritical CO_2 has been used for the extraction and isolation of valuable compounds from natural products.

Supercritical fluids are utilized to extract and purify natural colourant from eucalyptus bark (Vankar et al, 2001). Extraction of dye from food is best achieved with ethanol/oxalic acid. The comparative behaviour of other red food dyes is also studied and a process is developed for the extraction of natural dye from the leaves of teak plant is carried out using aqueous methanol (Nanda et al, 2001). A brick red shade from dyeing for silk/wool using the isolated dye in presence of different mordants is achieved. Attempts (Bhattacharya, 2002; Patel & Agarwal, 2001) has been made to standardize colourant derived from arjun bark, babool bark and pomegranate rind. Extraction (Agarwal et al 1992; Singh & Kaur, 2006) of well grounded henna leaves, directly in a solvent assisted dyeing process, employing organic solvent:water (1:9) as the dyeing medium is studied and superior dyeing properties are obtained, when applied to polyester. Natural dye (Raja & Kala, 2005) is obtained from the grape skin waste by using soxhlet extractor, and latter on distilled it under vacuum to obtain the concentrated dye solution. Colourant/dye is extracted by using a reflux condenser; source dye material is refluxed for 1 hour and filtered it to yield natural colourant (Eom et al, 2001).

5.3 Extraction by acid and alkali assisted system

Colour from euphorbia leaves (Dixit & Jahan, 2005) under acidic pH by adding hydrochloric acid and under alkaline pH by adding sodium carbonate both, in aqueous media are extracted for dyeing silk fabric. Extraction of colour in alkali media from nuts of acacia catechu (Sudhakar, 2006) is carried out for colouration of protein fibre based fabric. Dye extracted from jatropha seed gives a range of bright, even and soft colours on textiles when extracted under acid/alkali condition (Radhika & Jacob, 1999). Extraction of colour component from jackfruit wood under various pH conditions is carried out and reported that the optimum conditions for extraction is at pH-11.0 (Samanta et al, 2007). Red colour is extracted from red sandal wood (Samanta et al, 2006)under various pH conditions and it is reported that the optimum conditions for extraction of colour component is acidic pH like 4.0. Orangish yellow colour is extracted from tesu (palash flower) and maroon red colour can be obtained from Indian madder when extracted at alkaline conditions under aqueous medium (Samanta et al, 2010 & 2011).

5.4 Natural colour extraction by other methods

For ultrasound assisted extraction process (Sumate et al, 2008; Tiwari et al 2010) of natural colour, a standard extraction protocol may be used as 250 ml of solvent and 25 gm of powdered source natural dye material are taken in 500 ml beaker being immersed into the ultrasonic bath with working frequency of 27-30 MHz at 160 V under a controlled water level at about 2-3 cm from the bottom of the bath. To estimate the extraction yield at different time, temperature, pH and MLR for optimizing the extraction variables, 1.0 ml liquid is pipette out and then diluted to make 10.0 ml volume in each case. This solution is centrifuged at about 2000 rpm to remove the suspension. Finally, the concentration (% w/w) of the diluted solution is measured spectrophotometrically at a definite wavelength (λmax).

For enzyme assisted extraction process (Tiwari et al 2010) single or mixed enzyme (e.g. pectinase : cellulase, 2:1) is sprayed on source material and left for overnight for better soaking. This material is then taken into 500 ml conical flask with 250 ml water of pH 10 and shaken in orbital shaker at 150 rpm for 40-80 minutes at optimum temperature. The extraction solution is ready for dyeing textile material or can be filtered and drying as ready purified dye material for further use.

6. Purification and characterization of natural dyes

The aqueous extraction of the corresponding dye solution is double filtered in fine mesh nylon cloth and sintered glass crucible and the filtrate is evaporated using a vacuum oven at lower temperature (70°C) to a semi-dried solid mass and the same is then put in a cage of the wrapped filter paper and further subjected to extraction in soxhlet apparatus using 1:1 alcohol:toluene mixture for 10 cycles for 2h at 70°C. The alcohol- toluene extract of the colour components is finally subjected to evaporation in a water bath at 50°C to get a semi-dry mass of the pure colour components. Finally, this dry mass of the colour components is washed with 100% acetone followed by washing with methyl alcohol and final drying in air to obtain the dry powder of the pure colour components of the corresponding natural dyes.

For characterization, purified dye powder is to be taken for preparation of 1% aqueous dye solution separately and is subjected to wavelength scan in a micro processor or computer attached UV-Vis absorbance spectrophotometer for 190-1100 nm range. Further, individual purified natural dye powder is washed once again in distilled water and in 100% acetone in sequence before final drying and may be subjected to FTIR Spectroscopy study in double beam FTIR spectrophotometer using KBr disc technique for characterization of its chemical nature and functional group present in the natural dyes, which are responsible for solubilisation and mordanting power of natural dyes as well as its hypsochromic/bathochromic shift of the main hue.

For study of thermal behaviour by DSC (differential scanning calorimetry) study, individual purified natural dye powder is to be washed in distilled water followed by further washing in 100% acetone before final drying and then may be subjected to DSC or TGA (thermogravimetric analyser) study by standard method, for determining the different transition temperature of the purified dyes including temperature of degradation/dissociation. Thermal characterization by DSC/TGA is necessary for understanding the nature of thermal dissociation of natural dye component at different dyeing temperatures as well as application temperature.

6.1 UV-Visible spectroscopic study

UV-Vis spectral scan of aqueous/non-aqueous extract/solution of purified natural dyes having both UV-zone and visible zone (190-700 nm or higher) indicating peaks and troughs in different wave length shows its main hue, absorption etc. Peaks and troughs in visible zone thus indicate main colour and absorption. UV-Zone with/without peaks shows the property of the dye under UV-light, this may be correlated with fastness behaviour.

UV-Visible spectroscopic studies are carried out by different scientists (Erica t al, 1995) to identify the UV-Vis spectral scan of a number of natural dyes viz, madder, cochineal, indigo, etc., using different solvents for extraction. Neem bark (Mathur et al, 2003) colourant shows two absorption maxima at 275 and 374 nm while beet sugar shows three absorption bands at

220, 280 and 530 nm as per recent study (Mathur et al, 2001). The visible spectra of ratanjot (Gulrajani et al, 1999) at acidic pH showed maximum absorption around 520-525 nm, but under alkaline pH there is a shift to 570 nm and another peak at 610-615 nm and red sandal wood shows a strong absorption peak at 288 nm , the maximum absorption at 504 and 474 nm at pH 10 in methanol solution (Gulrajani et al, 2003). Gomphrena globosa (Sankar & Vankar, 2005) flower colourant shows one major peak at 533 nm. The dye does not show much difference in the visible spectrum at pH 4 and 7. Absorption for dyes extracted from mimusops elengi and terminalia arjun are reported that depending on the concentrations of dyes in the dye-bath, the dye absorbtion on the fibre varies from 21.94 % to 27.46 % and 5.18 % to 10.78% respectively (Bhuyan et al, 2004). The colour components isolated from most of the barks contain flavonoid moiety. Extraction, spectroscopic and colouring potential studies of the dye in ginger rhizome (zingiber officinale) is studied and reported (Popoola et al, 1995) that the dye is soluble in hydroxyl organic solvents and gives one homogenous component of R_f value of 0.86 on chromatographic separation having wavelength of maximum absorption at 420 nm. Aqueous extract of different source natural dyes including red sandal wood, manjistha, tesu, cutch etc. have been characterized by UV-vis spectra to optimize the extraction conditions.

6.2 Chromatographic analysis

Thin layer chromatatography (TLC) is used by many workers to identify natural dyes in textiles (Kharbade et al, 1985). Dyes detected are insect dyes and vegetable dyes viz., yellow, red and blue colours. The natural scale insect, madder and indigoid dyes are also analysed by HPLC (Koren, 1994). TLC chromatography analysis (Guinot et al, 2006) is used to carry out a preliminary evolution of plants containing flavonoids (flavonols, flavones, flavanones, chalcones/ aurones, anthocynanins), hydroxycinnamic acids, tannins and anthraquinones, which are the phylo-compounds (colour compounds) found in the plants. Identification of dyes in historic textiles through chromatographic and spectrophotometric methods as well as by sensitive colour reactions is highlighted (Blanc et al, 2006) and further the retention of carminic acid, indigotin, corcetin, gambogic acid, alizarin flavanoid, anthraquinone and purpurin are also studied (Szostek et al, 2003). A non-destructive method is reported for identifying faded dyes on textiles fabrics through examination of their emission and excitation spectra. The quantitative and qualitative analysis of red dyes such as alizarin, purpurin, carminic acid etc. by HPLC are also investigated/analysed (Balankina et al, 2006). High Performance Liquid Chromatography (HPLC) has been also used by several workers to identify natural dyes.

The separation and identification of natural dyes is carried out from wool fibres using reverse phase HPLC with a C-18 column (Mc Govern et al, 1990). Two quaternary solvent systems and one binary solvent system are used to obtain chromatograms of dyes, isomers and minor products present in the sample. A linear gradient elution method has been applied to the HPLC analysis of plant and scale insect, red anthraquinonoid, mordant dyes, and molluscan blue, red purple and indigoid vat dyes (Koren, 1994). The method enables the use of the same elution programme for the determination of different chemical classes of dyes. In addition, it significantly shortens the retention time of natural anthraquinonoid dyes. Quantitative analysis of weld by HPLC shows that after a 15 min extraction in a methanol-water mixture, 0.45% luteolin, 0.36 % luteolin 7-glucoside and 0.23% luteolin-3'7-diglucoside are obtained (Cristea et al, 2003). HPLC analysis of indigo is reported (Son et al, 2007) and it is found that as the dyeing time is increased, structural changes of indigo component are attributed to decrease in colour strength of dyeing.

Detection of annatto dyestuff, norbixin and bixin is reported by means of derivative spectroscopy and high performance liquid chromatography (HPLC) (Bhattacharya, 1999). The sample preparation involved extraction with acetone in the presence of HCl and removal of water by evaporation with ethanol. This residue is dissolved in chloroform-acetic acid for derivative spectroscopy or with acetone for HPLC. Derivative spectra are recorded from 550-400nm. Analysis of cochineal colour in foods utilizing methylation with diazomethane is carried out emplying TLC and HPLC using a mobile phase of butanol/ethanol and 10% acetic acid.

6.3 Test of toxicity, biotechnological processing and environmental impact of natural dyes

Toxicity is the ability of a substance to cause damage to living tissue, impairment of nervous system or severe illness when ingested, inhaled or being absorbed by skin. The toxicity (Zippel, 2004; Joshi & Purwar, 2004) data provide evidence about the adverse effect of natural dyes to human body. The LD_{50} is the best-known figure for toxicity rating of any substance. It describes the 'lethal dose for 50% of the test animals' which is the amount of substance in kg/kg of body weight which kills half of the animals.

Most of the natural dyes are found to be non-carcinogenic in nature. Moreover, natural dyes have positive effect on antifungal and anti bacterial growth. The crude methanolic extracts of stem and roots stem, leaves, fruit, seeds of artocarpus hetrophyllus (Khan et al, 2003) and their subsequent partitioning with petrol, dichloromethane, ethyl acetate and butanol fractions exhibit a broad spectrum of antibacterial activity. The butanol fractions of the root, bark and fruit are found to be the most active. None of the fraction is found to be active against the fungi tested. Mariegold (http://www.mdidea.com, 2005) shows negative test against microbiology control E-coli and salimonella. Chemo-preventative effects (Dwivedi & Ghazaleh, 1997; Dwivedi & Zhang, 1999; Benencia & Courreges, 1999) of red sandal wood oil are observed on skin papillomas in mice. Further, it is studied for red sandal wood's prevention of skin tumor development in CD1 mice and antiviral activity against herpes simplex virus-1 and 2. The hepatoprotective (Gilani & Janbaz, 1995) activity of an aqueous-methanol extract of rubia cardifolia (madder) is investigated against acetaminophen and CCL_4-induced damage. Acetaminophen produced 100% mortality at a dose of 1 g/K in mice while pretreatment of animals with plant extract (natural dye source material) reduced the death rate to 30%. Test of acacia catechu (cutch), kerria lacca (lac), quercus infectoria (gallnut), rubia cordifolia (madder) and rumex maritimus (golden dock) against pathogens like escherichia coli, bacillus subtilis, klebsiella pneumoniae, proteus vulgaris and pseudomonas aeruginosa are also reported (Singh et al, 2005). Minimum inhibitory concentration is found to be varying from 5 to 40 µg. Using a bioassay-directed purification scheme, the active antibacterial principle from caesalpina sappan (sappan wood or red wood) is isolated and identified (Hong & Lee, 2004). The trypan blue dye exclusion test shows that brazilian lacks cytotoxicity against vero cells; it has potential to be developed into an antibiotic. It is reported (Bhattacharya et al, 2004) that arjun bark, babool bark and pomegranate rind are eco-safe, however sometime contains traces amount of red listed heavy chemicals in permissible limit. A critical and realistic evaluation of dyeing with vegetable dyes highlighting its metal toxicity of substances used in processing has been reported (Shenai, 2002) and it is mentionworty that mordanting with metal salt as pre-requisite for application of most of the natural dyes may contaminate the dyed textiles with objectionable heavy metals resulting carcinogenic effect. Therefore, selection of mordanting metal salt and its purity are important criteria to produce eco-friendly natural dyed textiles. Attempts (Mondhe & Rao, 1993a & 1993b) has been made to prepare azo-alkyd dyes by the reduction of nitro alkyds, followed by diazotization of aminoalkyds and coupling with different phenolic compounds present in jatropha curcas seed oil confirmed by using IR spectra.

7. Types of mordants

Limitation on colour yield and poor fastness properties prompted a search for ideal mordants, the chemicals which increase natural dye uptake by textile fibres. Different types of mordants yield different colours even for the same natural dye. Therefore, final colour, their brilliance and colour fastness properties are not only dependant on the dye itself but are also determined by varying concentration and skillful manipulation of the mordants. Thus, a mordant is more important than the dye itself. Moreover, the ideal mordant for bulk use should produce appreciable colour yield in practicable dyeing conditions at low cost, without seriously affecting physical properties of fibre or fastness properties of the dyes. Also, It should not cause any noxious effect during processing and the dyed textile material should not have any carcinogenic effect during use. Mordants can be classified into the following categories:

7.1 Metallic mordants
They are generally metal salts of aluminium, chromium, iron, copper and tin. The metallic mordants are of two types.

7.1.1 Brightening mordants
i. **Alum:** Among all types of alum, potash alum is cheap, easily available and safe to use mordant. It usually produces pale versions of the prevailing dye colour in the plant.
ii. **Chrome (potassium dichromate):** It is also referred to as red chromate. It is relatively more expensive. However, Cr^{3+} or Cr^{6+} is considered to be harmful for human skin as objectionable heavy metal beyond a certain limit of its presence. Its use has been limited as per the norms of the eco-standards. The dichromate solution is light sensitive and therefore it changes colour under light exposure.
iii. **Tin (stannous chloride):** It gives brighter colours than any other mordant. However, they are oxidized on exposure to air and may impart a stiff hand to the fabric. Stannous chloride also causes higher loss of fabric tenacity (tensile strength) if applied beyond a certain concentrations.

7.1.2 Dulling mordants
i. **Copper (cupric sulphate):** Known as blue vitriol, it is readily soluble in water and easy to apply. It gives some special effects in shades, which otherwise cannot be obtained. However, copper beyond a certain limit is also under the eco-standard norms as objectionable heavy metals.
ii. **Iron (ferrous sulphate):** It is also known as green vitriol and is readily soluble in water. It is used for darkening /browning and blackening of the colours/ shades. It is easily available and one of the oldest mordants known. It is extensively used to get grey to black shades.

7.2 Tannins

The term 'tanning agent' is given initially to those water-soluble cellulosic materials that predicates gelatin from solution. But all gelatin precipitation did not identified as tanning agent. Tannins are polyphenolic compounds having capacity of gelling under certain conditions. (a) It may be hydroysable pyrogallol tannins exemplified by 'tannic acid', by Chinese or Turkish gallotannins (galls) and by Sicilain and Stagshorn sumac, (b) hydroysable ellagitannins that give ellagic acid or similar acids on hydrolysis, exemplified by valonea, chestnut, and (c) condensed or catechol tannins that contain little or no carbohydrates and are converted to acids to insoluble amorphous polymers. Among the tannins, myrobalan (harda) and galls/sumach are most important.

7.3 Oils type mordants

Vegetable oils or Turkey red oil (TRO) are such type of mordants. TRO as mordant is mainly used in the dyeing of deep red colour from madder. The main function of the TRO as oil mordant is to form a complex with alum when used as a main mordant. Sulphonated oil posses better binding-capacity than the natural oils. Oil mordanted samples exhibit superior fastness and hue.

8. Different mordanting methods and application of natural dyes

Mordanting can be achieved by pre-mordanting (before dyeing), simultaneously mordanting and dyeing or it may be a post mordanting system (after dyeing). Different types of mordants can be applied on the textile to increase the dye uptake of natural dyes. Extensive work has been reported (Paliwal, 2001; Jahan P & S, 2000; Sengupta, 2001; Prabu & Premraj, 2001; Sunita & Mahale, 2002; Moses, 2002; Rani & Singh, 2002; Bain et al, 2002; Paul et al, 2002) for dyeing of textiles with natural dyes adopting specific mordanting system for a particular textile material.

In pre-mordanting method, the textile substrate is first treated in aqueous solution of mordant for optimized time (e.g. 30 - 60 minutes) and temperature (e.g. 70 – 100 °C) with a ML ratio of 1:5 to 1:20 and then dried with or without washing. The mordanted textile material is then dyed following optimized dyeing conditions may be required as salt, soda ash or acid depending on type of textile material and type of natural dye. After dyeing, the textile material is washed properly and soaping is carried out by 2 g/L industrial soap solution as described in standard method of AATCC or ISO method.

For simultaneous mordanting and dyeing system, the textile substrate is immersed in a dye bath solution containing both mordant and dye in a definite quantity and dyeing may be started at the pre-determined optimum condition. Dyeing auxiliaries may be added as required for the standard dyeing process. However, for optimization of dyeing condition, dyeing process variables can be studied for specific fibre-mordant-natural dye system in order to maximize colour yield on textiles. After dyeing, the textile material is washed properly and soaping is carried out by 2 g/L industrial soap solution.

In case of post-mordanting method of natural dyeing, the dyeing process is carried out for bleached textiles in the absence of mordant at pre-determined dyeing condition and the dyed fabric is treated in a separate bath called saturator containing suitable mordanting solution. Treatment condition may vary depending on type of fibre, dye and mordant system. After dyeing, the textile material is washed properly and soaping is carried out by 2g/L industrial soap solution.

There is study (Dayal et al, 2006) for effect of copper sulphate and potassium dichromate on silk, wool and cotton fibre and reported their effects on colour fastness properties. The wool treated with metal ions such as Al(III), Cr (VI), Cu (II), Fe (II), Sn (II) and rare earths such as La (III), Sm (III) are used for beet sugar colourant, it can withstand the requirement of BIS fastness standards. Optimization (Agarwal et al, 1993) of the various concentrations of mordant are reported for shades can be produced by 0.15% of alum, 0.08% copper sulphate and stannous chloride, 0.04% ferrous sulphate and 0.06% potassium dichromate on mulberry silk fabric. Extraction of natural dye from the leaves of teak plant by using aqueous methonal produced brick red shade on dyeing of silk/ wool using the isolated dye in presence of different mordants as it is reported (Nanda et al, 2001). Irrespective of mordanting methods, silk (Mahale et al, 2003) treated with potash alum shows increase in colour when subject to sunlight test and those treated with potassium dichromate, copper sulphate and ferrous sulphate shows excellent to good fastness properties. Wool yarns dyed with turmeric (Mathur & Gupta, 2003) when subject to different concentration of natural mordant and chromium under identical mordanting conditions, shows similar colour fastness. Application of tulsi leave extract on textiles with or without using metallic salts produces pale to dark green and cream to brown shades with adequate fastness (Patel et al, 2002). Silk (Maulik & Pal, 2005) fabric being mordanted with magnesium sulphate produces lower depth of shade, whereas copper sulphate produces highest depth. It is reported (Bhattacharya & Shah, 2000) that the colour depth of dyeing textiles can be improved by using different metal salt as mordants. Pre-mordanting and post-mordanting (Das et al, 2005 & 2006) employing ferrous sulphate and aluminium sulphate improve the colour uptake, light fastness and colour retention on repeated washing for application of many natural dyes on textiles. The use of such mordants, however, does not improve wash fastness property of textile substrate dyed with pomegranate. Dyeing of wool (Chan et al, 2000) with four varieties of tea employing different mordant shows that coloured protein fibres became blackish, when ferrous sulphate is employed as mordanting agent. The effect of mordants on yellow dyes such as kapila, onion, tesu, and dolu are also reported.

Tin, as mordant imparts good wash fastness to cotton dyed with golden rods; chrome for mariegold dyeing and alum and tin for dyeing with onion skins (Vastard et al, 1999). Turmeric dye (Devi et al, 1999) can be applied for dyeing cotton fabric by using different mordants like tannic acid, alum, ferrous sulphate, stannous chloride and potassium dichromate to obtain various shades of colour. The use of gluconic acid as a ligand for complexing iron (II) salts and for vat dyeing of cotton has been studied. It is reported (Chavan & Chakraborty, 2001) for use of iron (II) salts complexed with ligands as tartaric acid and citric acid for the reduction of indigo at room temperature and subsequent cotton dyeing. Wash fastness (Kumar & Bharti, 1998) and light fastness (Sudhakar et al, 2006) can be increased by the use of metal salts or tannic acid on cotton fabrics. Cotton yarns treated with acalypha (Mahale, 2002) dye after pre-mordanted with potash alum, potassium dichromate, copper sulphate and ferrous sulphate shows excellent fastness rating.

Pre-mordanting route favours dyeing of jute (Samanta et al, 2003) fabric with direct type of natural dyes, when aluminium sulphate is used as a mordant, while simultaneous mordanting route gives better results for madder on cotton with the same mordant. It has also been proposed that alum (Potsch 1999) and aluminium sulphate should be used as mordants in dyeing with natural dyes, as their environmental toxicity is almost nil. The effects of different natural and chemical mordants like aluminium sulphate, tartaric acid and cetrimide on colour yield for bleached jute fabric are studied and reported (Samanta et al, 2006 & 2007). As the mordant concentration and dye concentration is increased, there is improvement in the light fastness by ½ to 1 grades. Different type of mordant and method of mordanting significantly affected the rate and extent of photofading. The use of copper or ferrous sulphate give high resistance to fading, whereas stannous chloride or alum did not. On the other hand, light fastness is improved when post-mordanting is conducted with copper or ferrous ion, but pre-mordanting is superior in the case of stannous chloride or alum as investigated and reported (Gupta et al, 2004). Harda-tartaric acid combination is found to be the best followed by tannic acid-harda and tartaric acid–tannic acid combinations. Synergistic effect of mordant is observed while using the binary combinations of mordants. Meta-mordanting gives the best results for harda-tartaric acid and tartaric acid-tannic acid combinations, while pre-mordanting gives the best results for tartaric acid-harda combination as studied (Deo & Paul, 2000a & 2000b). The colour fastness properties of goldendrop root dyed on wool (Bains et al, 2005) are studied using combinations of mordants such as alum: chrome, alum: copper sulphate, alum: ferrous sulphate, chrome: copper sulphate, chrome: ferrous sulphate and copper sulphate : ferrous sulphate in ratio of 1:3, 1:1 and 3:1. Studies are available for the effects of combination of mordant on colour fastness properties of cotton dyed with peach (Bains et al, 2003). There are lots of literatures available for mordanting prior to normal dyeing and the effects of mordants on colour fastness properties, shade development and other physical properties when applied singly (Fatima & Paul, 2005; Deo & Paul, 2003) or in combination (Yu et al, 2005) on cellulosic, protenic and synthetic fibres. An effective double pre-mordanting system is recommended (Samanta et al, 2006, 2007, 2010, 2011) for dyeing jute and cotton fabric using harda (as mordant assistant cum catcher) and aluminium sulphate (metallic mordant) without intermediate drying after mordanting, facilitating wet on wet dyeing in jigger. The myrobolan (harda) powder is soaked in water (1:10 volume) for overnight (12h) at room temperature to obtain the swelled myrobolan gel. This gel is then mixed with a known volume of water and heated at 80°C for 30 min. The solution is then cooled and filtered in a 60 mesh nylon cloth and the filtrate is used as final harda solution (10-40%) for 1st mordanting, using MLR of 1:20 (for dyed in bath) or 1:5 (for jigger). Pre-wetted conventional H2O2 bleached jute and cotton fabrics are separately treated with the harda solution in separate bath initially at 40-50°C and then the temperature is raised to 80°C. The mordanting is continued for 30 min. After the harda mordanting, fabric samples may be subjected to immediate wet on wet dyeing or may be dried in air without washing for storing purpose. Second mordanting is carried out using 10-40% of any one of the chemical mordants, (e.g., aluminium sulphate, potash alum, ferrous sulphate, stannous chloride and EDTA) at 80°C for 30 min using ML ratio of 1:20 (for dyed in bath) or 1:5 (for jigger). After the mordanting, the fabric samples may be finally dried in air without washing for storing purpose to make them ready for subsequent natural dyeing or may be subjected to immediate wet on wet dyeing without drying.

9. Principle of natural dyeing

Most of the natural dyes have no substantivity on cellulose or other textile fibres without the use of a mordant. The majority of natural dyes need a mordanting chemical (preferably metal salt or suitably coordinating complex forming agents) to create an affinity between the fibre and dye or the pigment molecules of natural colourant. These metallic salts as mordant form metal complexes with the fibres and the dyes. After mordanting, the metal salts anchoring to the fibres, attracts the dye/organic pigment molecules to be anchored to the fibres and finally creates the bridging link between the dye molecules and the fibre by forming coordinating complexes. Aluminium sulphate or other metallic mordants anchored to any fibre, chemically combine with certain mordantable functional groups present in the natural dyes and bound by coordinated/covelent bonds or hydrogen bonds and other interactional forces as shown below:

Mechanism of fixation of natural dyes through mordants

Thus, for proper fixation of natural dyes on any textile fibre, mordanting is essential in most of the cases. The said mordanting can be accomplished either before dyeing (pre-mordanting), or during dyeing (simultaneous mordanting) or after dyeing (post-mordanting).

9.1 Conventional methods of natural dyeing

Dyeing can be carried out in an alkaline bath, acidic bath or in a neutral bath. There are various reports available on different methods of mordanting on different fibers such as cellulosic, protenic and synthetic for dyeing with different natural dyes. Dyeing of cotton and silk with babool, tesu, manjistha, heena, indigo, mariegold etc is reported (Gulrajani et al, 1992; Saxena et al, 2001; Vankar et al, 2001; Nanda et al 2001; Patel & Agarwal, 2001). Various kinds of shades like black to brown, green to yellow to orange, etc can be obtained by application of different mordants.

9.1.1 Preparation of cotton fabric and dyeing with natural dyes

Cotton is purely cellulosic fibre and found throughout the world with many varieties and qualities. In general, cotton fibre based textiles is desized (for woven fabric only), scoured and bleached as preparatory process before dyeing with synthetic dyes. In many places of world, the age-old process followed in preparing a cotton cloth and its dyeing with natural dyes followed by artisan/cottage level dyers is given below (Mohanty et al, 1987):-

a. Dunging - The cloth is soaked for one night in a solution of water and fresh dung.

b. Washing - Next morning, cloth is thoroughly washed, rinsed and water sprinkling is continued over the cloth at short interval until evening, then it is finally washed and dried

c.   Steaming– Then the cloth is steamed for one night in an ordinary Khumb or washerman's steaming pot

d.   Steeping in alkaline lye- The cloth is soaked in a mixture of water, oil [castor oil or gingili oil], and alkali (sodium carbonate or soda known as sajikar or papadkhar).

e.   Rinsing- Cloth is then again rinsed thoroughly and spread out to dry.

f.   The last two processes are repeated for several days, the details varied in different localities, but generally from 3 to 7 days. In specific case, the cloth is kept in the solution for sometimes, and then taken out, rinsed and dried twice daily.

g.   Washing- The cloth is then finally washed in clean water, but not so thoroughly as to remove the whole of the oil, and finally dried in air under the sun.

h.   Galling-The cloth is then soaked in a solution of harda (haritaki) or myrobolan (Terminala chebula) extracts. Behda or bahedas (terminala belerica) is also used instead of harda. The period during which the cloth is kept in the harda extract varied in different places but it is continued until the fabric assume a yellowish tint.

i.   Drying- The cloth is spread or wrung out for drying.

j.   Mordanting- The cloth is then pre-mordanted by dipping it in a solution of potash alum and water. In some places, gum or a paste of tamarind seed (tamarind kernel powder) is added to make it sticky. In some parts of kutch, fuller's earth is also used by some dyers. The cloth is thus ready for subsequent dyeing.

k.   Dyeing- For dyeing the cloth is generally boiled with an aqueous extracted solution of the natural dye until all the colouring matter is absorbed by the cloth.

l.   Further dunging- In some places, the cloth is further soaked in dung for one night and batched before final wash and dry.

m.   Drying- The dyed fabric is next washed and spread out to dry gradually in air under the sun. Water is sprinkled at certain interval over the cloth, so as to brighten the colour, this process is continued for 2-4 days.

n.   Finishing- If required, the cloth is finally starched by dipping it in a paste of rice or wheat flour, or in a solution of babool gum and then dried.

However, now a days, many small scale dyers/export oriented units follow much shorter economical and standard recipe based optimized processes for natural dyeing of cotton yarns/fabrics. Before natural dyeing usual method of desizing (acid bath), scouring (soap & soda) and H_2O_2 bleaching are followed. Well prepared cotton textiles are then mordanted (single or double mordanting using harda and aluminium sulphate individually or in combination) before subjecting to dyeing with aqueous extract of selective natural dyes at standardized condition of process variables of dyeing. For e.g., the dyeing conditions may be as follows : dyeing time, 30 -120 minutes (depending on shades); dyeing temperature, 70-100°C; material to liquor ratio, 1:20 -1:30; concentration of natural dye, 10-50% (owm) or more; common salt concentration, 5-20g/L and pH, 10-12. In each case after the dyeing is over, the dyed samples are repeatedly washed with hot and cold water and then finally, the dyed samples are subjected to soaping with 2g/L soap solution at 60 °C for 15 min, followed by repeated water wash and line dried. For improving its wash fastness, treatment with eco-friendly cationic dye fixing agent is advisable.

9.1.2 Dyeing process for natural colouration of wool and silk fibre

Wool and silk are natural protein fibres and are available in wide variety having varied qualities. Both the fibres has complex chemical structure and very much susceptible to alkali attack (at pH >9). Hence, dyeing of these fibres with natural colours needs special care to avoid fibre damage by alkaline pH. Moreover, both wool and silk contain both amino and carboxylic functional groups. While, unlike silk, wool contains equal number of amino and carboxylic groups held together as salt linkages which bridge the main peptide chains. Therefore, in aqueous solution, wool carries no net charge. However, silk fibre has a slightly cationic character with the isoelectric point at about pH 5.0. Also, unlike wool fibre, silk is less sensitive to temperature. Therefore, selection of mordants, conditions of mordanting, pH and other conditions for dyeing, necessary cares are to be taken for colouration of these textile fibres for mordanting and application of natural dyes.

Wool and silk fibre based textiles can be dyed with different natural colours mostly through pre-mordanting or post mordanting system. Mordanting is done with tannin rich natural source chemicals like harda, gall nut etc and/or metal salts like, alum, aluminium sulphate, ferrous sulphate etc. Depending on shade depth requirement, mordant and dye concentration are to be determined. Dyeing conditions for a particular fibre-mordant–dye system need to be optimized by study of dyeing process variables before bulk dyeing.

In pre-mordanting system, these animal fibre based textiles are selectively mordanted (single or sequential double mordanted) with 5-20% (owm) mordant at 80-90°C for about 30-40 minutes having ML ratio 1:5-1:20 and can be taken for subsequent dyeing generally without washing. The pre-mordanted samples are entered in the dye bath (generally acidic dye bath) at 50-60°C and ML ratio 1:20, raise temperature upto the optimum dyeing temperature (may be 90°C) and dyeing is then continued for further 30 to 40 minutes followed by thorough rinsing, soaping and washing. However, in post mordanting system of applying natural colour on wool and silk textiles, dyeing is done at optimum dyeing condition and the dyed samples are dip into a mordant bath containing mainly 1-2% metal salt (owm) for true colour development followed by rinsing, soaping and thorough washing.

9.1.3 Dyeing process for natural colouration of jute and other lignocellulosic fibres

Application of natural colours on ligno-cellulosic fibres through double pre-mordanting system is found to be the best method, e.g., 10-20% harda treatment followed by 10-20% alum treatment or aluminium sulphate treatment is best suited mordanting system for subsequent dyeing of lingo-cellulosic fibre based textiles with different natural colours/dyes. Conventionally bleached and double pre-mordanted jute and linen textiles without wash, is dyed with tesu, madder, catechu, pomegranate rind, babool, jackfruit wood, haldi, marie gold, red sandal wood etc individually and in mixtures. Most of the above said natural dyes are applied in alkaline pH like 11-12 at higher temperature like 80-90°C having ML ratio 1:20 for 60 to 90 minutes. After dyeing, the dyed samples are washed and soaping is carried out at 60°C for 15 minutes. For getting higher fastness to wash and light, the samples can be further treated with 2% cationic dye fixing agents and/or 1% benztriazole.

9.1.4 Dyeing process for natural colouration of synthetic fibres

Different synthetic fibres like nylon, polyester etc are dyed with various source natural dyes/colourants like onion skin extract, babool bark extract, henna etc. through exhaust, HT-HP and padding methods (cold-pad-batch) with or without mordanting. Mordanting facilitates to get wide range of shades from the same source of natural dye as per requirement by variation of mordant chemicals as well as mordanting techniques. In case of dyeing synthetic fibres, dyeing is carried out at acidic pH and HT-HP dyeing technique results an overall best dyeing performance in terms of colour strength and fastness properties provided the natural colourants are stable at that high temperature employed for dyeing. Also energy consumption is higher in case of HT-HP dyeing method than exhaust method as well as cold-pad-batch method of dyeing.

Reports are available (Lokhande et al, 1998; Lokhande and Dorugade, 1999) for nylon is dyed with three different natural dyes using various mordants by two different techniques (open bath and high temperature high pressure dyeing methods), of which HT-HP dyeing is found better as compared to open bath. Application of babool bark extract on nylon substrate by cold-pad-batch and pad-dry-steam technique of dyeing can be considered as an effective eco-option and can be commercialized. HT-HP method is used for dyeing polyester fibre with pomegranate rind, catechu, nova red and turmeric (Bhattacharya and Lohiya, 2002).

9.2 Non-conventional dyeing methods

Nanotechnology is increasingly attracting worldwide attention because it is widely perceived as offering huge potential in a wide range of end uses. The unique and new properties of nanomaterials have attracted not only scientists and researchers but also businesses, due to their huge economical potential. One possible application is to directly employ pigment nanoparticles in textile coloration. Such an approach could be achieved if the nanoparticles can be reduced to a small enough size and the particles can be dispersed well to avoid aggregation of the nanoparticles in dye baths. Exhaust dyeing of cationized cotton with nanoscale pigment dispersion has recently been achieved and the results indicated that the dyeings obtained have better soft handle and more brilliant shade with reduced pigment requirement than those obtained with a conventional pigment dispersion (Fang et al, 2005).

Ultrasonic energized dyeing conditions for neem leaves gives better dye uptake, uniform dyeing, better light and wash fastness on cotton fabric (Senthikumar et al, 2002). Unconventional natural dyeing of cotton with sappan wood by ultrasound energy as well as new methods using microwave and sonicator for application of natural dye from alkanet root bark on cotton and dyeing of cotton fabrics with tulsi leaves extract by use of ultrasonic energy dyeing are also reported (Ghorpade et al, 200; Tiwari et al, 2000a & 2000b).

The application of supercritical carbon dioxide ($scCO_2$) in the textile industry has recently become an alternative technology for developing a more environmentally friendly coloration process. $scCO_2$ coloration technology has the potential to overcome several environmental and technical issues in many commercial textile applications such as yarn preparation, coloration and finishing. $scCO_2$ represent a potentially unique media for either transporting chemical into or out of a polymeric substrate, because of their thermo-physical and transport properties. Supercritical fluids exhibit gas-like viscosities and diffusivities and liquid-like densities. Additionally, carbon dioxide is nontoxic, non-flammable, environmentally friendly, and chemically inert under many conditions; however, its production is remained to be cost-effective. The dissolving power of $scCO_2$ for disperse dyes and its use as the transport media for coloration polyester is studied from all theoretical aspects at DTNW in Krefeld, Germany. Studies have revealed that the presence of intramolecular hydrogen bonds and/or the hydrophobicity of dye molecule are positive factors for better solubility in supercritical carbon dioxide, as indicated from its improved dye-uptake (Shakra et al, 1999 & 2000)

The use of $scCO_2$ as a fluid medium for coloration of textile fibers, especially polyester, has been examined. This technology has become so promising that it has provided new opportunities to develop suitable dyes for this medium. The coloration is conducted in a stainless steel high pressure apparatus. Process and equipment are developed for textile dyeing in supercritical carbon dioxide (Kraan, 2007). A technical-scale, 100-L dyeing machine is designed and built for polyester beam dyeing in $scCO_2$ at 300 bar.

10. Physico-chemical studies on dyeing process variables and dyeing kinetics

The natural dyes have a variable chemical composition, which is influenced by a number of factors, out of which the most important are: the vegetal part of the plant where the extract is obtained from, its place and growing conditions, harvesting period, extraction operation and application or technological process followed. Many workers (Gupta, 1999) have reported some of the most significant experimental results, laboratory trials regarding techniques and best parameters for extraction and application of natural dyes, i.e. observations on varying extraction parameters, such as: extraction temperature, extraction time, extraction solvent-vegetal material ratio, type of solvent and observations on varying mordanting and dyeing parameters. However, most of such studies are concentrated on wool and silk.

The effects of dye-extraction medium, optimum concentrations of dye source material, extraction time, dyeing time, mordant concentration and methods of mordanting on silk dyed with natural dyes has been reported (Grover et al, 2005; Dixit & Jahan, 2005; Dumitrescu et al, 2005; Srivastava et al, 2006; Das et al, 2005; Agarwal et al, 1992; Bansal & Sood, 2001; Sati et al, 2003; Rose et al, 2005; Maulik & Bhowmik, 2006; Siddiqui et al, 2006). The acidic media exhibited maximum percent absorption for jatropa, lantana, hamelia and euphorbia dye, while kilmora and walnut showed good results in alkaline medium. The result obtained from different experiments lead to the optimization of a standard recipe for particular dye-mordant-fibre combination. The optimum concentration of beet sugar (Mathur et al, 2003) colourant for dyeing wool is found to be 0.03g per g of wool at pH 4.5 and temperature 97.5oC. Dyeing of wool under the optimum (Mathur et al, 2001) condition pH 4.5, colourant concentration-0.05g per gram of wool; time-60 min and treatment temperature- 97.5oC shows very good light and wash fastness properties without deteriorating the quality of wool. Optimum dyeing technique for colouration of wool by determining the optimum wavelength, dye material concentration, extraction time, dyeing time, pH, concentration of mordant etc. helps to standardize the dyeing process (Srivastava et al, 2006). Colouring component of tea shows highest affinity for both wool and silk at pH 2 to 4 in presence and absence of ferrous sulphate and aluminium sulphate as mordants (Das et al, 2005). Optimisation of dyeing process variables for wool with natural dyes obtained from turmeric has been studied and reported (Agarwal et al, 1992). The optimum conditions for development of vegetable dye on cotton from eupatorium leaves are studied and reported (Bansal & Sood, 2001). The optimization of wool by using rhododendron arboretum as a natural dye source is reported (Sati et al, 2003). The effect of process variables on colour yield and colour fastness properties for application of selective natural dyes for different textiles are also studied and reported (Rose et al, 2005; Maulik & Bhowmik, 2006; Siddiqui et al, 2006).

The dyeing absorption isotherm for wool, human hair, silk, nylon and polyester is found to be linear indicating a partition mechanism of dyeing for application of juglone as natural dye (Gupta & Gulrajani, 1993). The slope of isotherms increases with the increasing temperature in all cases. ΔH and ΔS values are positive for all the dyeings. The apparent diffusion coefficient is highest for wool and lowest for silk. For investigation of the conditions of extraction and application of african mariegold on silk yarn, optimum conditions are found to be 60 minutes dye extraction time, 30 minutes mordanting, 30 minutes dyeing using mixtures of 5% potash alum, 1% potassium dichromate and 1% copper sulphate as mordants (Mahale et al, 1999). Studies of dyeing absorption isotherm, heat of dyeing, free energy and entropy of dyeing for red sandal wood (Samanta et al, 2006) and jackfruit wood (Samanta & Agarwal, 2008) reveals that both the dyeing process follows a linear Nernst absorption isotherm. An adsorption and thermodynamic study of lac dyeing on cotton pre-treated with chitosan showed Langmuir isotherm (Rattanaphani et al, 2007). The colour can vary significantly for indigo dyeing on denim yarns as the result of the variation of parameters (Kin et al, 2007).

11. Compatibility of mixture of selective natural dyes

Newer shades can be achieved by applying mixture of natural dyes. For the use of mixture of natural dyes, the dyers must know whether the natural dyes are compatible with each other or not. For test of compatibility of a pair of natural dyes, bleached and pre-mordanted samples are dyed in two different sets of progressive depth of shade of binary mixture of natural dyes as follows:-

Bleached and pre-mordanted textiles are dyed with the specific binary mixtures of natural dye pair. The dyeing is started at 40°C and the temperature is gradually raised to 100°C taking total time of 60 min (approximately) to raise this temperature at a heating rate of around 1°C/min, using precision temperature controller open bath Laboratory beaker-dyeing machine.

In set I, (progressive depth of shade developed by varying dyeing time and temperature profile during dyeing), for 1% shade with binary mixture (50:50) of a pair of dyes, pre-mordanted samples are separately dyed for different dyeing period (10, 20, 30, 40, 50 and 60 min), by withdrawing separately dyed samples from the dye bath at the intervals of 10 min from 50°C onwards maintaining the heating rate of around 1°C/min. The penultimate sample is taken out after 50-60 min at 90°C and the last one at the end of the dyeing carried out for 60 min at 100°C.

In set II (progressive depth of shade developed by varying total concentrations of dye mixture using 20-100 parts of 1% shade depth using purified natural dye-stuff colourants) for a pair of binary mixture dyes, pre-mordanted samples are separately dyed at increment of twenty percentage points by applying 20 -100 parts of 1% dye (on weight of fabric) for each pair of natural dye-mixture taken in equal proportion (50:50) at 100°C for 60 min. For both Set I and Set II, after dyeing, all the dyed samples are subjected to normal washing, soaping, and rinsing before final air-drying. The corresponding surface colour strength (K/S value) and the differences in the CIELAB coordinates namely, ΔL, Δa, Δb and ΔC for all dyed fabric samples for Set I and II obtained indicate for the samples dyed with using purified natural dyes, the colour yield (K/S), lightness/darkness (ΔL), redness/greenness (Δa), yellowness/blueness (Δb) and differences in saturation/chroma (ΔC) values with respect to the standard un-dyed sample, which are measured and obtained from separate measurement of the same using the reflectance spectrophotometer and associated software and computer. Plots of K/S vs ΔL and/or ΔC vs ΔL, i.e., two sets of curves obtained for the said two sets (Set-I & Set-II) of dyed samples indicate the nature of compatibility by closeness of the pattern of the two sets of curves (Konar, 2011). However, another easy method of Relative Compatibility Rating (RCR) method (Samanta et al, 2009) has been established based on differences between lowest and highest colour difference index (CDI) values [CDI = (ΔE X ΔH) / (ΔC X MI)] for a binary mixture of natural dyes dyed in different proportions with fixed shade % at standardized dyeing conditions.

12. Colour fastness properties of natural dyes

Colour fastness is the resistance of a material to change in any of its colour characteristics or extent of transfer of its colourants to adjacent white materials in touch or both for different environmental and use conditions or treatments like washing, dry cleaning etc or exposure to different agency heat, light etc. Fading means changes in the colour with or without loss of depth of shade for exposure to particular environment/agency/treatments either by lightening or darkening of the shades. Bleeding is the transfer of colour to a secondary material in contact accompanying white fibre material of similar/dissimilar nature. The colour fastness is usually rated either by loss of depth of colour/ colour change in original sample or it is often expressed by staining scale meaning that the accompanying material gets tinted/stained by the colour of the original fabric, when the accompanying white fabrics of similar/dissimilar nature are either in touch/ made to touch by some means of test procedure/protocol.

12.1 Light fastness

An extensive work has been carried out to improve the light fastness properties of natural dyed textiles. A comprehensive review on different attempts taken for improving colour fastness properties of dyes on different textile fibres by different means is reported (Cook, 1982). The said review includes tannin-related after-treatments for improving the wash fastness and light fastness of mordantable dyes on cotton; some of these treatments might be applicable to selective/specific natural dyes.

Most of the natural dyes have poor light stability (as compared to that of the best synthetic dyes), and hence the colours in museum textile are often different from their original colours. The relative light stability of a range of dyes has been reviewed (Padfield & Landi, 1966) along with studies involving change in qualitative fashion. These changes in colour are studied quantitatively (Duff et al, 1977) where it is expressed the changes in terms of the Munsell scale and also in CIE colour parameters. Wool dyed with nine natural dyes is exposed in Microscal MBTF fading lamp. The fastness ratings are similar to those found earlier in day light fading. After rating by the blue wool standards for light fastness rating, yellow dyes (old fustic, persian berries) shows poor light fastness between 1-2; red colours like cochineal(tin mordant), alizarin (alum and tin mordant), lac (tin mordant) shows better light fastness between 3-4 ; indigo shows light fastness 3-4 or 5-6 (depending on the mordant) ; and logwood black (chrome mordant) shows light fastness 4-5 or 6-7 (other mordants). It is also reported the effects of chemical structure of natural dyes on light fastness and other colour fastness properties (Gupta, 1999a & 1999b).

A large proportion of natural dyes are, of course, mordant dyes. There is strong influence on nature, type and concentrations of mordants on wash and light fastness grades. The influences of different mordants are found to play important role in fading of 18 yellow natural dyes (Crews, 1982). Where wool dyed with different natural dyes specimens are exposed to a xenon arc lamp for assessing their light fastness upto 8 AATCC Fading Units equivalent to BS-8B Blue wool standards. The corresponding colors changes after exposure to xenon arc lamp are also assessed in each case.

Turmeric, fustic and marigold dyes faded significantly more than any of the other yellow dyes. However, use of tin and alum mordants resulted in significantly more fading than the same for use of chrome, iron, or copper mordant. Thus the type of mordant is found to be is more important than the dye itself in determining the light fastness of natural colored textiles.

Natural dyes that may have been used in the traditional Scottish textile industry have been described (Grierson, 1984) that the light fastness of such dyes on wool has been compared with those of dyeing with 'imported' dyes to similar shades (Duff et al, 1985; Grierson et al,1985), again using Microscal fading MBTF lamp.

Numerous attempts (Samanta et al, 2006, 2010 & 2011; Hofenk, 1983; Oda, 2001; Cristea & Vilarem, 2006; Lee, 2001; Micheal & Zaher, 2005; Gupta, 2001) has been made to improve the light fastness of different textiles fabric dyed with natural dyes which include the effects of various additives on the photofading of carthamin in cellulose acetate film, critical examination of fading process of natural dyes with a view to determining the original colours of faded textile etc. The rate of photofading is remarkably suppressed in the presence of nickel hydroxy-arylsulphonates, while the addition of UV absorbers afforded little retardation in the rate of fading.

12.2 Wash fastness

With a view to examine and improving wash fastness (Duff et al, 1977), tests are carried out under standard condition (50ºC) and also at 20ºC with a washing formulation used in conservation work for restoration of old textiles. Some dyes undergo marked changes in hue on washing, shown to be attributed to even small amounts of alkali in washing mixtures, high-lighting the necessity of knowing the pH of alkaline solutions used for cleaning of textiles dyed with natural dyes. As a general rule, natural dyes (on wool) have only moderate wash fastness as assessed by the ISO 2 test. However, logwood and indigo dyes exibit better fastness when applied to different textiles. The nature of detergent solution suitable for conservation of natural coloured art work has been examined (Hofenk, 1983). A liquor containing 1g/l of sodium polyphosphate is found to be best resulting marginal changes in hue with natural dyes applied on wool or silk (Duff et al, 1977). The small increase in cleaning efficiency attributable to the alkali must be balanced against possible colour change in the natural dyes, apart from possible damage to the protein fibre under alkaline conditions.

In the ISO 2 test, the fastness of the indigo and logwood is superior to that of the native natural dyeing such as privet berries and water lily root respectively, but in the comparison of native and imported yellow, reds, red/purples, greens and browns, there is little difference between the two groups (Duff et al, 1977). It is found from a recent report that treatment with 2%CTAB or sandofix-HCF improve the wash fastness to nearly 1 unit and treatment with 1% benztriazole improved the lightfastness of dyed jute textiles nearly half to one unit (Samanta et al, 2006; 2007; 2010 & 2011; Samanta & Agarwal, 2008)

12.3 Rubfastness

Rub fastness of most of the natural dyes have been found to be moderate to good and dose not require any after treatment. Jackfruit wood, manjistha, red sandal wood, babool, mariegold etc have good rubfastness (Samanta et al, 2006; 2007; 2010 & 2011). Good rub fastness is seen for mariegold on cotton, silk and wool (Sarkar et al, 2005; Sarkar, 2006). Good rub fastness (dry and wet rubfastness) is reported for silk dyed with acalypha and other natural dyes (Mahale et al, 1999; 2002 & 2003). Cutch and ratanjot shows moderate to good dry rub fastness but the wet rub fastness is found to be average (Khan et al, 2003 & 2006).

13. Concluding remark

Most of the natural dyes/ colour are eco-safe, except a few. Some of the natural colours are not only eco-safe, but also has added value for its medicinal effects on skin and are more than skin friendly. Textile dyers must know the chemistry of these natural colours and its added advantages of medicinal; values. Use of suitable binary opr ternary mixtures of similar or compatible natural dyes for colouring natural eco-friendly textiles in variety of soothing / uncommon shades with eco-friendly mordants and finishing agents are the most desireable product of the customers for future. The non-reproducibility and poor colour fastness etc, have been partly solved by many researchers' continuous effeorts in this endevour. So, a textile dyer must know the effects of variability for extraction, mordanting and dyeing and should follow only the standardized recipe for selection fibre-mordant-natural dye system to get reproducible colour yield and colour matching besides to follow different eco-friendly ways to improve colour fastness to a possible extent. This chapter is a highlight of all these efforts towards popularizing natural dyeing not only in small/cottage dyers but also textile yarn/fabrics industrially using common dyeing machine. Some of the study mentioned above explained this clearly. So, natural dyeing/colouration of textiles by industrial processes in large scale dyeing unit is now a reality in the textile market of eco-friendly textiles.

14. References

Agarwal A, Goel A & Gupta K C, *Textile Dyers and Printer*, 25(10), (1992), 28

Agarwal A, Garg A & Gupta K C, *Colourage*, 39(10) (1992) 43.

Agarwal A, Paul S and Gupta K C , *Indian Text J*, (1) (1993) 110.

Bansal S & Sood A, Textile Magazine, 42 (8), (2001), 81

Bhattacharya S K, Dutta C & Chatterjee S M, *Man-made Textiles in India*, (8) (2002) 207.

Blanc R, Espejo T, Montes A L, Toress D, Crovetto G, Navalon A & Vilchez J L, *J Chromatography A*, 1122 (2006) 105

Balankina G G, Vasiliev V G, Karpova E V & V I Mamatyuk, *Dyes Pigm*, 71(2006) 54.

Bhattacharya N, in Proceeding of convention of Natural Dyes edited by Deepti Gupta & M.L Gulrajani, Department of Textile Technology, IIT Delhi, 1999, 134

Benencia F & Courreges M C, *Phytomed*, 6 (2) (1999) 119.

Bhattacharya S K, Chatterjee S M & Dutta C , *Man-made textiles in India*, (2004) 85

Bain S , Singh O P & Kang K , *Man-made textiles in India*, 45(8) (2002) 315.

Bhattacharya S D & Shah A K , *J Soc Dyers Color*, 116 (1) (2000) 10.

Bains S , Kaur K & Kang S , *Colourage*, 52 (5) (2005) 51.

Bains S, Kang S & Kaur K , *Man-made Textiles in India*, 46(6) (2003) 230.

Bhuyan R, Saikai C N &Das K K, *Indian J Fibre Text Res*, 29(12) (2004)429, 470.

Bhattacharya N and Lohiya N, *Asian Textile J*, 11(1), (2002), 70

Cristea D, Bareau I & Vailarem G, *Dyes Pigm*, 57 (2003) 267.

Chan P M , Yuen C W M & Yeung K W , *Textile Asia*, 31(2) (2000) 28.

Chavan R B & Chakraborty J N, *Colouration Technol*, (2001), 117.

Cook C C , *Rev. Prog. Colouration*, 12 (1982) 78-89

Crews P C, *J American Institute Conservation*, 21 (1982) 43-58.

Cristea D & Vilarem G, *Dyes and Pigments* , 70, (2006), 238

Duff D G , Sinclair R S & Grierson S , *Textile History*, 16 (1) (1985) 23-43

Duff D G, Sinclair R S & Stiriling D , *Studies in Conservation* 22 (1977) 161-169

Dedhia E M, *Colourage*, 45 (3), 1998, 45

Dayal R & Dobhal P C , *Colourage*, 48 (8) (2001) 33

Dixit S & Jahan S , *Man-made Textiles in India*, 48 (7) (2005) 252.

Dwivedi C & Ghazaleh A A, *Eur J Cancer Prev*, 6(4) (1997) 399.

Dwivedi C & Zhang Y, *Eur J Cancer Prev*, 8 (5) (1999) 449.

Dayal R , Dhobal P C , Kumar R , Onial P & Rawat R D, *Colourage*, 53(12) (2006) 53.

Devi W A , Gogoi A, Khanikar D P, *Indian Text* J,109 (12) (1999) 60.

Deo H T & Paul R, *Indian J Fibre Text Res*, 25(6) (2000a) 152.

Deo H T & Paul R, *Indian J Fibre Text Res*, 25(9) (2000b) 217.

Deo H T & Paul R , *International dyers*, 188 (11) (2003) 49.

Das D , Bhattacharya S C & Maulik S R, *Indian J Fibre Text Res*, 31 (12) (2006) 559

Das D , Bhattacharya S C & Maulik S R, *International J Tea Science*,4 (3 & 4) (2005) 17.

Dumitrescu I, Mocioiu A I & Mocioiu A M, 56(4), (2005), 235, 2456, *World Text Abst*, 38 (4), (2006), 299

Erica J, Tiedemann and Yiqi Yang, *J Am Inst Conserv*, 34, (3) (1995) 195

Eom S , Shin D & Yoon K , *Indian J Fibre Text Res*, 26 (12) (2001) 425.

Fang K, Wang C, Zhang X, Xu Y (2005) Color Technol 121:325

Fatima N & Paul S, *International dyers*, 190 (2) (2005) 24.

Gulrajani M L & Gupta D, Natural dyes and application to textiles, Department of textile technology, Indian Institute of Technology, New Delhi, India, 1992

Gulrajani M L , Gupta D & Maulik S R, *Indian J Fibre Text Res*, 24 (1999) 294.

Gulrajani M L, Bhaumik S, Oppermann W & Hardtmann G, *Indian J Fibre Text Res*, 28 (2003) 221.

Guinot P , Roge A, Argadennec A, Garcia M, Dupont D, Lecoeur E, Candelier L & Andary C, Colouration Technology, (122) (2006) 93.

Gilani A H & Janbaz K H , *Phytotherapy Res*, 9 (5) (1995) 372.

Gupta D , Gulrajani M L & Kumari S *Colouration Technol*, 120 (2004) 205.

Ghorpade B, Darrekar M and Vankar P S, *Colourage*, 47(1), (2000), 27

Gupta G, *in Proceedings of 1st Convention on Natural Dyes* (Ed-D Gupta & M.L Gulrajani), Dept of Textiles Tech, IIT Delhi, 9th –11th December, (1999), 121

Grover E, Sharma A & Rawat B, *International dyers*, 190(10), (2005), 9

Gupta D P, Gulrajani M L, *Indian J Fibre Text Res*, 18 (12), (1993), 202

Gupta D , *Colourage*, 46 (7), (1999a), 35 & 46 (8), (1999b), 41

Grierson S, *J Soc Dyers Color*, 100 (1984) 209-211.

Grierson S , Duff D G & Sinclair R S , *J Soc Dyers Color*, 101 (1985)

Hong-Xi Xu & Lee Song F, *Phytotherapy Res,* 18 (2004) 647.

http://www.mdidea.com/products/herbextract/mariegold/data.html (29/8/2005)

Hofenk J H de Graaff, in *'Conservation-Restoration of Church Textiles and Painted Flags'*, 4th Int Restorer Seminar, Hungary (1983), Vol. 2, 219-228.

Jahan P & Jahan S , *Indian Silk,* 40 (9) (2000) 31.

Joshi M & Purwar R, *Rev Prog Color,* 34 (2004) 58.

Kumar V & Bharti B V , *Indian Text J,* (2) (1998) 18

Khan M A , Khan M , Srivastav P K & Mohammad F, *Colourage,* 56 (1) (2006) 61.

Konar A, PhD Thesis, *'Studies on Textile Related Properties and Dyeability of Jute and Chemically Modified Jute Textiles using Selective Synthetic and Natural Dyes'* Jadavpur University, 2011

Kharbade B V & Agarwal O P , J *Chromatography,* 347 (1985) 447.

Koren Z C, *J Soc Dyers Color,* 110 (9) (1994) 273.

Khan M R , Omoloso A D & Kihara M, *Fitoterapia,* 74 (5) (2003) 501.

Kraan Van Der, PhD Thesis: http://repository.tudelft.nl/file/82909/027985. August, 2007

Kin J H, Chong C L & Tu T, *J Soc Dyers Color,* 116 (9), (2000), 260

Lokhande H T, Dorugade V A and Naik S R, *American Dyestuff Reporter,*87(6), (1998), 40

Lokhande H T and Dorugade V A, *American Dyestuff Reporter,* 88 (2), (1999), 29

Lee J J, Lee H H, Eom S I & Kim J P , *Colouration Technology,* 117, (2001), 134

Maulik S R & Pradhan S C, *Man-made Textiles in India,* 48 (10) (2005) 396.

Mathur J P , Metha A , Kanawar R & Bhandaru C S , *Indian J Fibre Text Res,* 28 (2003) 94.

Mathur J P & Bhandari C S , *Indian J Fibre Text Res,* 26 (2001) 313.

Mc Govern P.E, Lazar J & Michel R H , *J Soc Dyers Color,* 106 (1) (1990) 22.

Mondhe O P S & Rao J T , *Colourage,* 43(5) (1993a) 43.

Mondhe O P S & Rao J T , *Colourage,* 43(6) (1993b) 51.

Moses J J , *Asian Texs J,* 11(7) (2002) 68.

Mahale G, Sakshi & Sunanda R K , *Indian J Fibre Text Res,* 28(3) (2003) 86

Mathur J P & Gupta N P, *Indian J Fibre Text Res,* 28 (2003) 90.

Maulik S R & Pal P , *Man-made Textiles in India,* 48 (1) (2005) 19.

Mahale G, Sakshi & Sunanda R K, *International Dyers,* 187 (9) (2002) 39.

Mohanty B C, Chandramouli K V & Naik H D, *Natural dyeing processes of India,* published by Calico Museum of Textiles, Ahmedabad, India, 1987

Maulik S R & Bhowmik L, Man-made textiles in India,49(4), (2006), 142

Mahale G, Bhavani K, Sunanda R K & Sakshi M, *Man-made textiles in India,*42(11), (1999), 453

Micheal M N & Zaher N A El, *Colourage,* 2005), (Annual) 83

Nanda B, Nayak A, Das N B, Patra S K, in Proceeding of convention of Natural Dyes edited by Deepti Gupta & M.L Gulrajani, Department of Textile Technology, IIT Delhi, 2001, 85.

Oda H , *Colouration Technology,* 117 (4), (2001), 204 & 117 (5), (2001), 254

Padfield P & Landi S, *Studies in Conservation,* 11 (1966), 161-196

Patel R & Agarwal B J, in Proceeding of convention of Natural Dyes edited by Deepti Gupta & M.L Gulrajani, Department of Textile Technology, IIT Delhi , 2001,167.

Pan N C, Chattopadhyay S N & Day A, *Indian J Fibre Text Res,* 28 (9) (2003) 339.

Popoola A V, Ipinmoroti K O, Adetuyi A O & Ogunmoroti T O, *Pakistan J of Scientific and Industrial Research*, 37(5), 1994, 217, 95W/05090, *World Text Abstr*, 27(8) (1995) 452.

Paliwal J, *Textile Magazine*, 42 (11) (2001) 79.

Prabu H G & Premraj L , *Man-made Textiles in India*, 44 (12) (2001) 488.

Paul S, Sharma A & Grover E, *Asian Text J*, 11(11) (2002) 65.

Patel K J , Patel B H , Naik J A & Bhavar A M , *Man-made Textiles in India*, (11) 2002.

Potsch W R, *Melland Textiberichte*, 80(11-12) 1999, 967, 2480, *World Text Abstr*, 32(4),2000,282

Rani A & Singh O P, *Asian Text J,* 11(9) (2002) 47.

Raja N Vasugi & Kala J, *J Text Assoc,* 66 (3) (2005) 117.

Rattanaphani S, Chairat M , Rattanaphani J B, *Dyes and pigments*, 72 (1), (2007), 88

Rose N M, Khanna S, Singh J S S & Gabba G, Textile Trend, 48(4), (2005), 45

Radhika D & Jacob M , *Indian Text J,* 109 (7) (1999) 30.

Sudhakar R, Ninge K N & Padaki N V, *Colourage*, 53 (7) (2006) 61.

Samanta A K, Agarwal P, Singhee D & Datta S, *J Text Inst*, 100(7) (2009) 565

Samanta A K, Agarwal P & Datta S, *Indian J Fibre & Text Res*, 32 (12) (2007) 466.

Samanta A K, Agarwal P & Datta S, *J Inst Engg (I), Text Engg;* 87 (2006) 16.

Samanta A K, Konar A, Chakroborty S & Datta S, *Indian J Fibre & Text Res*, 36 (3) (2011) 63

Samanta A K, Konar A, Chakroborty S & Datta S, *J Inst Engg (I), Text Engg;* 91 (2010) 7.

Sumate Boonkird T J, Phisalaphong C & Phisalaphong M, Ultrasonics Sonochem, 15, (2008) 1075

Saxena S, Varadarajan P V & Nachane N D, in Proceeding of convention of Natural Dyes edited by Deepti Gupta & M.L Gulrajani, Department of Textile Technology, IIT Delhi, 2001, 185.

Sarkar D, Mazumdar K, Datta S & Sinha D K, *J Textile Assoc*, 66 (2) (2005) 67.

Sarkar D, Mazumdar K & Datta S, *Man-made Textiles in India*, (1) (2006) 19.

Singh K & Kaur V, *Colourage*, 53 (10) (2006) 60.

Sankar R & Vankar P S, *Colourage*, 52 (4) (2005) 35.

Szostek S, Grwrys J O, Surowiec I & Trojanowicz M, *J Chromatography A*, 1012 (2003) 179.

Son A-Y, Hong P J & Kim K T, *Dyes Pigm*, 61(3) (2007) 63.

Singh R, Jain A , Panwar S , Gupta D & Khare S K, *Dyes Pigm* ,66(2) (2005) 99.

Shenai V A, *Colourage*, 49 (10) (2002) 29.

Sengupta S, *J Text Assoc*, 62(4) (2001) 161.

Sunita M B & Mahale G , *Man-made Textiles in India*, 45(5) (2002) 198.

Sudhakar K N, Gowda N & Padaki N V , *Colourage*, 53 (7) (2006) 61.

Samanta A K, Singhee D, Sengupta A & Rahim A S, *J Inst Engg(I), Text Engg;* 83 (2) (2003) 22.

Senthikumar S, Umashankar P and Sujatha B, *Indian Textile J*, 112(6), (2002), 15

Shakra S, Mousa AA, Youssef BM, El-kharadly EA (1999) Mans Sci Bull (A Chem) 26(2):1

Shakra S, Mousa AA, Youssef BM, El-kharadly EA (2000) Mans Sci Bull (A Chem) 27(2):1

Srivastava M, Pareek M & Valentina, *Colourage*, 53(2), (2006), 57

Sati O P, Rawat U & Srivastav B, Colourage, 50(12), (2003), 43

Siddiqui I, Gous Md & Khaleq Md S, Indian Silk, 45(4), (2006), 17

Samanta A K & Agarwal P, *Indian J Fibre & Text Res*, 33 (3), (2008), 66

Tiwari V, B Ghorpade and P S Vankar, *Colourage*, 47 (3), (2000a), 21

Tiwari V, B Ghorpade, A Mishra and P S Vankar, *New Cloth Market*, 14 (1), (2000b), 23

Teli M D & Paul R , *International Dyers*, 191 (4) (2006) 29.

Tiwari H C, Singh P, Mishra P K & Srivastava P, *Indian J Fibre & Text Res*, 35 (9) (2010) 272

Vankar P S, Tiwari V & Ghorpade B, in Proceeding of convention of Natural Dyes edited by Deepti Gupta & M.L Gulrajani, Department of Textile Technology, IIT Delhi, 2001, 53.

Verma N & Gupta N P, *Colourage*, 42 (7) (1995) 27.

Vastard J, Shailaja D & Mamatha A, *Indian Text J*, 109 (7) (1999) 68.

Yu B , Wu Q & Yu L, *International dyers*, 190 (5) (2005) 23.

Zippel E , *Rev Prog Color*, 34 (2004) 1.

3

Dyeing in Computer Graphics

Yuki Morimoto^{1,3}, Kenji Ono^{1,2} and Daisaku Arita^3
^{1}*RIKEN*
^{2}*The University of Tokyo*
^{3}*ISIT (Institute of Systems, Information Technologies and Nanotechnologies)*
Japan

1. Introduction

In this chapter, we introduce a physically-based framework for visual simulation of dyeing. Since ancient times, dyeing has been employed to color fabrics in both industry and arts and crafts. Various dyeing techniques are practiced throughout the world, such as wax-resist dyeing (*batik dyeing*), hand drawing with dye and paste (*Yuzen dyeing*), and many other techniques Polakoff (1971); Yoshiko (2002). Tie-dyeing produces beautiful and unique dyed patterns. The tie-dyeing process involves performing various geometric operations (folding, stitching, tying, clamping, pressing, etc.) on a support medium, then dipping the medium into a dyebath. The process of dipping a cloth into a dyebath is called dip dyeing.

The design of dye patterns is complicated by factors such as dye transfer and cloth deformation. Professional dyers predict final dye patterns based on heuristics; they tap into years of experience and intimate knowledge of traditional dyeing techniques. Furthermore, the real dyeing process is time-consuming. For example, clamp resist dyeing requires the dyer to fashion wooden templates to press the cloth during dyeing. Templates used in this technique can be very complex. Hand dyed patterns require the dyer's experience, skill, and effort, which are combined with the chemical and physical properties of the materials. This allows the dyer to generate interesting and unique patterns. There are no other painting techniques that are associated with the deformation of the support medium. In contrast to hand dyeing, dyeing simulation allow for an inexpensive, fast, and accessible way to create dyed patterns. We focus on dye transfer phenomenon and woven folded cloth geometry as important factors to model dyed patterns. Some characteristic features of liquid diffusion on cloth that are influenced by weave patterns, such as thin spots and mottles are shown in Figure 1. Also, we adopted some typical models of adsorption isotherms to simply show adsorption. Figure 2 shows the simulated results obtained using our physics-based dyeing framework and a real dyed pattern. Figure 3 depicts the framework with a corresponding dyeing process.

2. Related work

Non-photorealistic rendering (NPR) methods for painting and transferring pigments on paper have been developed for watercolor and Chinese ink paintings Bruce (2001); Chu & Tai (2005); Curtis et al. (1997); Kunii et al. (2001); Wilson (2004). These methods are often based on fluid

Fig. 1. Some characteristic features of dyeing. (a) thin spots, (b) bleeding ("nijimi" in Japanese), and (c) mottles.

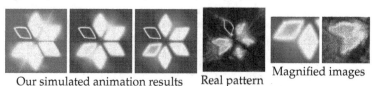

Our simulated animation results   Real pattern   Magnified images

Fig. 2. Comparisons of our simulated results with a real dyed pattern.

| (a) Define weave (Plain weave) | (b) Specify cloth geometry | | (c) Supply dyes (Dip dyeing) |
| | Fold | Press (Stitch) | |

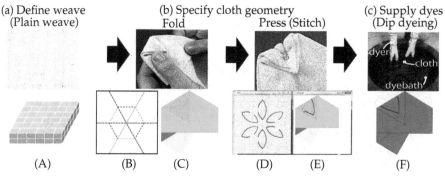

(A)   (B)   (C)   (D)   (E)   (F)

Fig. 3. The general steps for a real dyeing process (top row Sakakibara (1999)) and our dyeing framework (bottom row) using the Chinese flower resist technique. (A) The woven cloth has a plain weave; the blue and yellow cells indicate the warp and the weft. (B) The unfolded cloth with user specified fold lines, the red and blue lines indicate ridges and valleys, and folds in the cloth. (C) The corresponding folded cloth in (B). (D, E) The interfaces representing user drawings on an unfolded and folded cloth. The gray lines indicate a user-specified boundary domain; these will be the dye resist regions. (F) The folded cloth with the red region indicating the exterior surfaces.

mechanics: Kunii et al. (2001) used Fick's second law of diffusion to describe water spreading on dry paper with pigments; Curtis et al. (1997) developed a technique for simulating watercolor effects; Chu & Tai (2005) presented a real-time drawing algorithm based on the lattice Boltzmann equation ; and Xu et al. (2007) proposed a generic pigment model based on rules and formulations derived from studies of adsorption and diffusion in surface chemistry and the textile industry.

Several studies have also investigated dyeing methods: Wyvill et al. (2004) proposed an algorithm for rendering cracks in *batik* that is capable of producing realistic crack patterns in wax; Shamey (2005) used a numerical simulation of the dyebath to study mass transfer in a fluid influenced by dispersion; and Morimoto et al. (2007) used a diffusion model that includes adsorption to reproduce the details of dyeing characteristics such as thin colored threads by considering the woven cloth, based on dye physics.

However, the above methods are insufficient for simulating advanced dyeing techniques, such as tie-dyeing. Previous methods are strictly 2D, and are not designed to handle the folded 3D geometry of the support medium. There is no other simulation method that considers the folded 3D geometry of the support medium which clearly affects real dyeing results except for Morimoto et al. (2010). Morimoto et al. (2010) proposed a simulation framework that simulates dyed patterns produced by folding the cloth. In this chapter, we summerize these studies.

3. Cloth weave model

We represent the cloth by cells (see Figure 4). We define two kinds of cells, namely a cloth cell, and a diffusion cell. In cloth cells, parameters such as the weft or warp, the vertical position relative to the weave patterns, the size of threads and gaps, and other physical diffusion factors specific for each thread are defined. Each cloth cell is subdivided into several diffusion cells and these diffusion cells are then used to calculate the diffusion.

Initially, cloth cells are defined by the size of the warps and wefts, which can be assigned arbitrarily. They are then arranged in the cloth at intervals equal to the spacing of gaps. Two layers are prepared as the weft and warp for cloth cells. Next, a defined weave pattern determines whether each cloth cell is orientated up or down. Diffusion cells are also arranged in the cloth size and two layers are formed. Their properties are defined in reference to the cloth cell in the layer that they are in. Each diffusion cell can have only one of two possible orientations (parallel to the x and y axes) as the fiber's orientation in its layer. Gravity is considered to be negligible in our dye transfer model since the dye is usually a colloidal liquid; this enables us to calculate the diffusion without considering the height in the cloth. There is thus no necessity to arrange diffusion cells in three-dimensional space, making it is easy to consider the connection between threads. The effect of height is considered in our model only for cloth rendering.

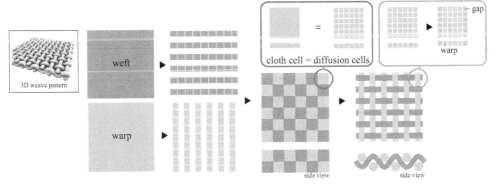

Fig. 4. The example of cloth cells with plain weave.

4. Dye transfer model

Our dye transfer model accounts for the diffusion, adsorption, and supply terms of the dye as described in an equation (1). The diffusion and adsorption terms describe the dye behavior based on the second law of Fick (1855) and dyeing physics, respectively. The supply term enables arbitrary dye distribution for dip dyeing and user drawings.

Let $f = f(\mathbf{x}, t) \in (0, 1]$ be a dye concentration function with position vector $\mathbf{x} \in \Re^3$ and time parameter $t \geq 0$, where t is independent of $\mathbf{x}$. The dyeing model is formulated by the following evolutionary system of PDEs.

$$\frac{\partial f(\mathbf{x}, t)}{\partial t} = \text{div}(D(\mathbf{x})\nabla f) + s(\mathbf{x}, f) - a(\mathbf{x}, f), \tag{1}$$

where $D(\mathbf{x})$ is the diffusion coefficient function, $\text{div}(\cdot)$ and $\nabla(\cdot)$ are the divergence and gradient operators respectively, and $s(\cdot, \cdot)$ and $a(\cdot, \cdot)$ are the source and sink terms that represent dye supply and adsorption, whereas $\text{div}(D(\mathbf{x})\nabla f)$ is the diffusion term. We model the functions $s(\cdot, \cdot)$ and $a(\cdot, \cdot)$ as follows.

$$s(\mathbf{x}, f) = \begin{cases} \alpha M_s(\mathbf{x}) & \text{if } M_s(\mathbf{x}) > f(\mathbf{x}, t) \text{ and } M_{cd}(\mathbf{x}) > f(\mathbf{x}, t). \\ 0 & \text{Otherwise,} \end{cases}$$

$$a(\mathbf{x}, f) = \begin{cases} \beta f(\mathbf{x}, t) & \text{if } h(\mathbf{x}, t) < a_d(\mathbf{x}, f) \text{ and } M_{ca}(\mathbf{x}) > h(\mathbf{x}, t). \\ 0 & \text{Otherwise,} \end{cases}$$

where α is the user-specified dye concentration, $M_s(\mathbf{x}) \in [0, 1]$ is the dye supply map, $M_{cd}(\mathbf{x})$ and $M_{ca}(\mathbf{x})$ are the diffusion and adsorption capacity maps, $\beta \in [0, 1]$ is the user-specified adsorption rate, and $a_d(\mathbf{x}, f)$ is the adsorption capacity according to adsorption isotherms. The adsorption isotherm depicts the amount of adsorbate on the adsorbent as a function of its concentration at constant temperature. In this model, we employ the Langmuir adsorption model Langmuir (1916), which is a saturation curve, to calculate $a_d(\mathbf{x}, f)$ in our simulation based on the model developed by Morimoto et al. Morimoto et al. (2007). The adsorbed dye concentration $h(\mathbf{x}, t) \in [0, 1]$ is given by the equation bellow and the evolution of $f(\mathbf{x}, t)$ and $h(\mathbf{x}, t)$ as $t \to \infty$ describes the dyeing process.

$$\frac{\partial h(\mathbf{x}, t)}{\partial t} = a(\mathbf{x}, f)\frac{M_{cd}(\mathbf{x})}{M_{ca}(\mathbf{x})}, \tag{2}$$

Designing physically-based diffusion coefficient.

Each diffusion cell has various parameters including its weft layer or warp layer, fiber or gap (i.e., no fiber), orientation (either up or down), position, porosity, and tortuosity. Porosity is defined as the ratio of void volume in a thread fiber. Tortuosity denotes the degree of twist, so that the smaller the value of the tortuosity is, the larger the twist is. We define three kinds of tortuosities in our method:

- the twist of the thread ($\tau_1(\mathbf{x})$) defined in each thread

- the position of the thread, including its orientation in the weave pattern ($\tau_2(\mathbf{x})$) determined by the connection between neighboring diffusion cells, whether they are in the same layer, whether they contain fibers, and whether their orientation is up or down.

- the different orientations of fibers in neighboring diffusion cells ($\tau_3(\mathbf{x})$)

Each of these tortuosities has values in the range (0, 1]. There are five different conditions for $\tau_3(\mathbf{x})$ which depend on factors such as the layers that neighboring diffusion cells are located in and their porosity. In Figure 5, colored lines indicate the conditions such as bellow list.

I (blue): Different layer

II (orange): Two fivers are in the same layer, and are connected to each other perpendicularly

III (purple): Fiber and gap

IV (green): gap and gap

V (red): fiver and fiver in the same layer

In Figure 5, blue, yellow and white cells represent wefts, warps and gaps. Finally, the tortuosity of a cell $T(\mathbf{x})$ between diffusion cells is defined as follows:

$$T(\mathbf{x}) = \tau_1(\mathbf{x})\tau_2(\mathbf{x})\tau_3(\mathbf{x}) \tag{3}$$

The diffusion coefficient is calculated between diffusion cells using the following equation that is based on the Weisz-Zollinger modelS.-H. (August 1997) in accordance with our definition for $T(\mathbf{x})$:

$$D(\mathbf{x}) = D_0\mathbf{p}(\mathbf{x})T(\mathbf{x})f_0 \tag{4}$$

where $\mathbf{p}$ is the porosity that can have any value in the range (0, 1], f_0 denotes the dye concentration in the external solution when equilibrium is achieved and is specified arbitrarily. D_0 is the diffusion coefficient in free water and is calculated according to the following equation: Van den (31 Deccember 1994)

$$D_0 = 3.6\sqrt{76/M} \tag{5}$$

where M denotes the molecular mass of the dye.

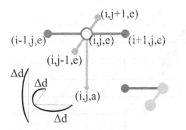

Fig. 5. Connection between diffusion cells with tortuosity. The third factor in parentheses indicate the warp / weft layer (a / e).

Boundary condition.

Let $\mathbf{z} \in \mathcal{B}$ be the boundary domain for dye transfer. We use the Neumann boundary condition $\frac{\partial f(\mathbf{z},t)}{\partial \mathbf{n}(\mathbf{z})} = b(\mathbf{z}) = 0, \mathbf{z} \in \mathcal{B}$ where $\mathbf{n}(\mathbf{z})$ is a unit normal vector of the boundary. In tie-dyeing, parts of the cloth are pressed together by folding and pressing. We assume that the pressed region is the boundary domain because no space exists between the pressed cloth parts for the dye to enter. In our framework, the user specifies $\mathcal{B}$ by drawing on both the unfolded and folded cloth, as shown in Figure 3 (D) and (E), respectively. In the case of the folded cloth, shown in Figure 3 (E), we project $\mathcal{B}$ to all overlapping faces of the folded cloth. Here, the faces are the polygons of the folded cloth, as shown in the paragraph, Diffusion graph construction.

Press function.

We introduce a press function $P(\mathbf{x}, c) \in [0, 1]$ as a press effect from dyeing technique. Here c is a user-specified cut-off constant describing the domain of influence of the dye supply and the capacity maps, and represents the physical parameters. The extent of the pressing effect and dye permeation depend on the both softness and elasticity of the cloth and on the tying strength. The press function serves to;

- Limit dye supply (Pressed regions prevent dye diffusion).
- Decrease the dye capacity (Pressed regions have low porosity).

We approximate the magnitude of pressure using the distance field $dist(\mathbf{x}, \mathcal{B})$ obtained from the pressed boundary domain $\mathcal{B}$. Note that the press effect only influences the interior surfaces of the cloth, as only interior surfaces can press each other.

We define Ω as the set of exterior surfaces of the tied cloth that are in contact with the dye and $\mathcal{L}$ as the fold lines that the user input to specify the folds. We define Ω, $\mathcal{B}$, and $\mathcal{L}$ as fold features (Figure 6). We calculate the press function $P(\mathbf{x}, c)$ as described in Pseudocode 1. In Pseudocodes 1 and 2, $CalcDF()$ calculates the distance field obtained from the fold feature indicated by the first argument, and returns infinity on the fold features indicated by the second and third arguments. We use m as the number of vertices in the diffusion graph described in the paragraph "Diffusion graph construction."

Dye supply map.

The dye supply map $M_s(\mathbf{x})$ (Figure 7) describes the distribution of dye sources and sinks on the cloth, and is applied to the dye supply term in equation (1). In dip dyeing, the dye is supplied to both the exterior and interior surfaces of the folded cloth. The folded cloth opens naturally except in pressed regions, allowing liquid to enter the spaces between the folds (Figure 8). Thus, we assume that $M_s(\mathbf{x})$ is inversely proportional to the distance from Ω, as it is easier to expose regions that are closer to the liquid dye. Also, the dye supply range depends on the movement of the cloth in the dyebath. We model this effect by limiting $dist(\mathbf{x}, \Omega)$ to a cut-off constant c_Ω. Another cut-off constant $c_{\mathcal{B}_1}$ limits the influence of the press function $P(\mathbf{x}, c_{\mathcal{B}_1})$.

The method used to model the dye supply map $M_s(\mathbf{x})$ is similar to the method used to model the press function, and is described in Pseudocode 2. In Pseudocode 2, $dist_{max}$ is the max value of $dist(\mathbf{x}, \Omega)$ and $GaussFilter()$ is a Gaussian function used to mimic rounded-edge folding (as opposed to sharp-edge folding) of the cloth.

$dist(\mathbf{x}, \mathcal{B}) \leftarrow CalcDF(\mathcal{B}, \Omega, \mathcal{L})$
for i=1 to m **do**
  if $dist(i, \mathcal{B}) > c$ **then**
    $P(i, c) \leftarrow 1.0$
  else
    $P(i, c) \leftarrow dist(i, \mathcal{B})/c$
  end if
end for

Pseudocode 1. Press function.

$dist(\mathbf{x}, \Omega) \leftarrow CalcDF(\Omega, \mathcal{B}, \mathcal{L})$
for i=1 to m **do**
  $dist(i, \Omega) \leftarrow (dist_{max} - dist(i, \Omega))$
  if $dist(i, \Omega) > c_\Omega$ **then**
    $dist(i, \Omega) \leftarrow P(i, c_{\mathcal{B}_1})$
  else
    $dist(i, \Omega) \leftarrow P(i, c_{\mathcal{B}_1})dist(i, \Omega)/c_\Omega$
  end if
end for
$M_s(\mathbf{x}) \leftarrow GaussFilter(dist(\mathbf{x}, \Omega))$

Pseudocode 2. Dye supply map.

Capacity maps.

The capacity maps indicate the dye capacities in a cloth; they define the spaces that dyes can occupy. We first calculate the basic capacities from a fibre porosity parameter Morimoto et al. (2007). The cut-off constant $c_{\mathcal{B}_2}$ limits the press function $P(\mathbf{x}, c_{\mathcal{B}_2})$ used here. We then multiply these capacities by the press function to obtain the capacity maps $M_{cd}(\mathbf{x})$ and $M_{ca}(\mathbf{x})$ as followings (see Figure 9).

$$M_{cd}(\mathbf{x}) = V_d(\mathbf{x})P(\mathbf{x}, c_{\mathcal{B}_2}), \qquad\qquad M_{ca}(\mathbf{x}) = V_a(\mathbf{x})P(\mathbf{x}, c_{\mathcal{B}_2}), \qquad (6)$$

The basic capacity of the dye absorption in the diffusion cell $V_a(\mathbf{x})$ is determined by the ratio of fiber in the diffusion cell according to the following expression:

$$V_a(\mathbf{x}) = 1 - \mathbf{p}(\mathbf{x}) \qquad (7)$$

Also, the maximum amount of dye adsorption (fixing dye) into fibers $(a_d(\mathbf{x}, f))$ is calculated each time step by adsorption isotherm based on the dyeing theories used in our model Langmuir (1916), Vickerstaff (1954). Figure 13 shows representative behavior of these typical adsorption isotherms. If we use the Freundlich equation (Figure 13 (b")) in our system, we can define $a_d(\mathbf{x}, f)$ as:

$$a_d = kf^b \qquad (8)$$

where k is a constant, and $\mathcal{B}$ is a constant that lies in the range (0.1, 1). If $b = 1$, eq 8 become a linear equilibrium as shown in Figure 13 (a"). If we use the Langmuir equation (Figure 13 (c"))

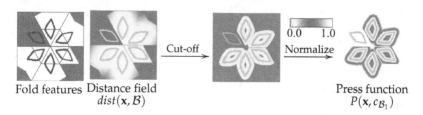

Fold features | Distance field $dist(\mathbf{x}, \mathcal{B})$ | Cut-off → | | Normalize → | Press function $P(\mathbf{x}, c_{\mathcal{B}_1})$

Fig. 6. An example for calculating the press function $P(\mathbf{x}, c_{\mathcal{B}_1})$ for the Chinese flower resist. In the fold features, the exterior surface Ω, pressed boundary domain $\mathcal{B}$, and the fold lines $\mathcal{L}$ are shown by red, gray, and blue colors.

Distance field $dist(\mathbf{x}, \Omega)$

Invert, cut-off by c_Ω, normalize →

Multiply by $P(\mathbf{x}, c_{\mathcal{B}_1})$ →

Smoothing →

Dye supply map $M_s(\mathbf{x})$

Fig. 7. An illustration of the dye supply map $M_s(\mathbf{x})$ for the Chinese flower resist.

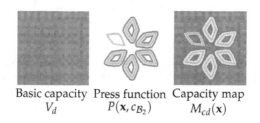

Fig. 8. Photographs of real dip dyeing. The left shows a folded cloth pressed between two wooden plates. The right shows folded cloths that have opened naturally in liquid.

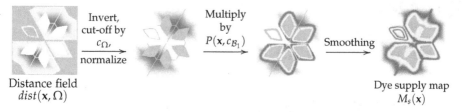

Basic capacity V_d | Press function $P(\mathbf{x}, c_{\mathcal{B}_2})$ | Capacity map $M_{cd}(\mathbf{x})$

Fig. 9. Illustration for our diffusion capacity map $M_{cd}(\mathbf{x})$ calculated by taking the product of the basic capacity and $P(\mathbf{x}, c_{\mathcal{B}_2})$.

in our model, we can defined $a_d(\mathbf{x}, f)$ as:

$$a_d(\mathbf{x}, f) = \frac{M_{cd}(\mathbf{x})K_L f}{1 + K_L f} \tag{9}$$

where K_L is the equilibrium constant. The resulting images are shown in Figure 12 and are calculated using these adsorption isotherms. The parameter $b(\mathbf{z})$ prevents dyeing in the pressed area of the cloth and is used to represent the dyeing patterns of certain dyeing techniques. Curtis et al. (1997) used a similar method in which the paper affects fluid flow to some extent for watercolor and by doing so generates patterns. But the effect of this phenomenon is not as pronounced as that of dyeing patterns. In our method, we use the amount of dye that is not absorbed by the fiber to calculate the total amount of dye in the diffusion cell. The basic diffusion capacity $V_d(\mathbf{x})$ is calculated as bellow,

$$V_d(\mathbf{x}) = \mathbf{p}(\mathbf{x})(1 - b(\mathbf{z})), \tag{10}$$

Diffusion graph construction.

We construct multiple two-layered cells from the two-layered cellular cloth model by cloth-folding operations. The diffusion graph $\mathcal{G}$ is a weighted 3D graph with vertices $\mathbf{v}_i$, edges, and weights w_{ij} where $i, j = 1, 2, .., m$. The vertex $\mathbf{v}_i$ in the diffusion graph $\mathcal{G}$ is given by the cell center.

We apply the ORIPA algorithm Mitani (2008), which generates the folded paper geometry from the development diagram, to the user-specified fold lines on a rectangular cloth. The fold lines divide the cloth into a set of faces as shown in Figure 3 (B). The ORIPA algorithm generates the corresponding vertex positions of faces on the folded and unfolded cloths and the overlapping relation between every two faces as shown in Figure 3 (C). We apply ID rendering to the faces to obtain the overlapping relation for multiple two-layered cells. We then determine the contact areas between cells in the folded cloth, and construct the diffusion graph $\mathcal{G}$ by connecting all vertices $\mathbf{v}_i$ to vertices $\mathbf{v}_j$ in the adjacent contact cell by edges, as illustrated in Figure 10.

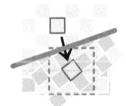

(a) An illustration for the multiple two-layered cloth model.

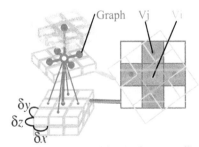

(b) A graph in dot square in (a). (c) Contact cells.

Fig. 10. Graph construction from folded geometry of woven cloth. The bold line in (a) represents a fold line. The gray points in (c) are contact cell vertices $\mathbf{v}_j$ of the target cell vertex $\mathbf{v}_i$.

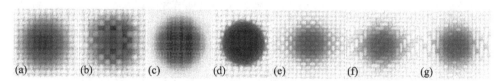

(a)   (b)   (c)   (d)   (e)   (f)   (g)

Fig. 11. Computer generated dye stains with various parameters including $T(\mathbf{x}), \mathbf{p}(\mathbf{x})$, weave, etc. Total number of time steps is 5000.

Discretization in diffusion graph.

The finite difference approximation of the diffusion term of equation (1) at vertex $\mathbf{v}_i$ of $\mathcal{G}$ is given by

$$\text{div}D(\mathbf{x})\nabla f \approx \sum_{j \in N(i)} w_{ij} \frac{f(\mathbf{v}_j, t) - f(\mathbf{v}_i, t)}{|\mathbf{v}_j - \mathbf{v}_i|^2}, \tag{11}$$

where $w_{ij} = D_{ij} A_{ij}$, $N(i)$ is an index set of vertices adjacent to $\mathbf{v}_i$ in $\mathcal{G}$, and D_{ij} and A_{ij} are the diffusion coefficient and the contact area ratio between vertices $\mathbf{v}_i$ and $\mathbf{v}_j$, respectively (Figure 10). We calculate the distance between $\mathbf{v}_i$ and $\mathbf{v}_j$ from cell sizes, $\delta x, \delta y, \delta z$, and define $D_{ij} = (D(\mathbf{v}_i) + D(\mathbf{v}_j))/2$ between vertices connected by folding.

The following semi-implicit scheme gives our discrete formulation of equation (1):

$$(I - (\delta t)L)\mathbf{f}^{n+1} = \mathbf{f}^n + \delta t(\mathbf{s}^n - \mathbf{a}^n), \tag{12}$$

which is solved using the SOR solver Press et al. (1992), where I is the identity matrix, δt is the discrete time-step parameter, n is the time step, and

$$\mathbf{f}^n = \{f(\mathbf{v}_1, n), f(\mathbf{v}_2, n), .., f(\mathbf{v}_m, n)\},$$

$$\mathbf{s}^n = \{s(\mathbf{v}_1, f(\mathbf{v}_1, n)), s(\mathbf{v}_2, f(\mathbf{v}_2, n)), .., s(\mathbf{v}_m, f(\mathbf{v}_m, n))\},$$

$$\mathbf{a}^n = \{a(\mathbf{v}_1, f(\mathbf{v}_1, n)), a(\mathbf{v}_2, f(\mathbf{v}_2, n)), .., a(\mathbf{v}_m, f(\mathbf{v}_m, n))\},$$

L is the $m \times m$ graph Laplacian matrix Chung (1997) of $\mathcal{G}$. The element l_{ij} of L is given by

$$l_{ij} = \begin{cases} w_{ij}/|\mathbf{v}_j - \mathbf{v}_i|^2 & \text{if } i \neq j, \\ -\sum_{j \in N(i)} l_{ij} & \text{Otherwise.} \end{cases} \tag{13}$$

We then simply apply the forward Euler scheme to equation (2):

$$h(\mathbf{v}_i, n+1) = h(\mathbf{v}_i, n) + \delta t a(\mathbf{v}_i, f(\mathbf{v}_i, n)) \frac{M_{cd}(\mathbf{v}_i)}{M_{ca}(\mathbf{v}_i)} \tag{14}$$

The diffused and adsorbed dye amounts at $\mathbf{v}_i$ are given by $f(\mathbf{v}_i, n)M_{cd}(\mathbf{v}_i)$ and $h(\mathbf{v}_i, n)M_{ca}(\mathbf{v}_i)$, respectively. The dye transfer calculation stops when the evolution of equation (1)converges to $\sum_{i=1}^m |f(\mathbf{v}_i, n)^n - f(\mathbf{v}_i, n)^{n-1}|/m \leq \epsilon$.

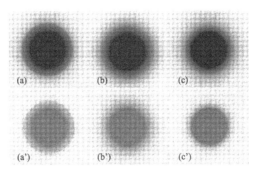

Fig. 12. Comparison of simulations with some adsorption isotherms. Total numbers of time steps is 50000.

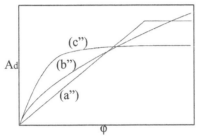

Fig. 13. Behavior of some typical adsorption isotherms.

5. Result

For visualizations of the dyed cloth, we render the images by taking the product of the sum of dye (transferred and adsorbed) and its corresponding weft and warp texture.

Physically-based model

The results of our simulation are shown in Figs. 11, 12. Figure 11 (d) shows mottles similar to those in Figure 1 (c). This shows that mottles are formed with not just when the value of $\tau_3 I(\mathbf{x})$ is small but they are also produced by the two-layered model. Figures 11 (f) and (g) show the results for random $\mathbf{p}(\mathbf{x})$, while Figure 11 (e) shows the result for random $\tau_1(\mathbf{x})$. Some thin colored threads can clearly be seen in Figure 11 (f). We find that $\mathbf{p}(\mathbf{x})$ has more effect on the final appearance than $\tau_1(\mathbf{x})$. Figure 11 (b) shows the result obtained when a high absorption coefficient is used; in this result, the initial dyeing area seems to contain a lot of dye.

Figure 12 shows results calculated using some adsorption isotherms as shown in Figure 13. (a) and (b) are the result obtained using the Freundlich model (eq 8) where $b = 1, k = 1$ and $b = 0.3, k = 1$. (c) is the result obtained using the Langmuir model (eq 9). (a'), (b') and (c') show only the absorbed dye densities of (a), (b) and (c), respectively. In this comparison, we can observe the effect of the kind of adsorption isotherm has on the results obtained.

Folded cloth geometry

Figure 14 shows real dyed cloth and our simulated results for a selection of tie-dyeing techniques. The dyeing results are evaluated by examining the color gradation, which stems from cloth geometry. Our framework is capable of generating heterogeneous dyeing results by visualizing the dye transferring process. While we cannot perform a direct comparison of our simulated results to real results, as we are unable to precisely match the initial conditions or account for other detailed factors in the dyeing process, our simulated results (a, b) corrspond well with real dyeing results (c).

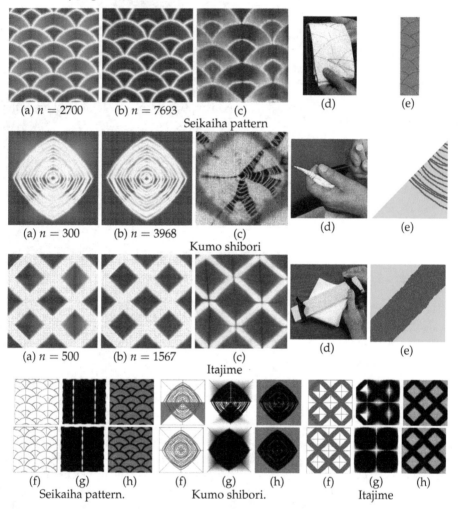

(a) $n = 2700$ (b) $n = 7693$ (c)   (d)   (e)
Seikaiha pattern

(a) $n = 300$ (b) $n = 3968$ (c)   (d)   (e)
Kumo shibori

(a) $n = 500$ (b) $n = 1567$ (c)   (d)   (e)
Itajime

(f)   (g)   (h)   (f)   (g)   (h)   (f)   (g)   (h)
Seikaiha pattern.   Kumo shibori.   Itajime

Fig. 14. Various tie-dyeing results and their corresponding conditions. (a) Our simulated results. (b) Our converged simulated results. (c) Real tie-dyeing results Sakakibara (1999). (d) Tie-dyeing techniques. (e) Folded cloths with user-specified boundary domains. (f) Fold features. (g) Dye supply maps $M_s(\mathbf{x})$. (h) Dye capacity maps $M_{cd}(\mathbf{x})$. In (f), (g), and (h), the top and bottom images indicate the top and bottom layers of the cloth model, respectively. The thin gray regions in the Seikaiha pattern (e) indicate regions that were covered with a plastic sheet to prevent dye supply on them; they are not the boundary domain. n is the number of time steps.

6. Conclusion

In this article, we introduced physically-based dyeing simulation frameworks. The framework is able to generate a wide range of dyed patterns produced using a folded woven cloth, which are difficult to produce by conventional methods.

It is important to create real-world dyed stuffs using this system as an user study. However, these systems are not enough to simulate complex dyed patterns intuitively. For example, an interface for modeling dyeing techniques such as complex cloth geometries and coloring are important but they are not achieved. Other future work to simulate dyed patterns by PC is enriching physical simulation of dyeing. In the CG dyeing methods we introduced, there is no advection and color mixing. These factors make the dyeing system attractive and useful more.

After these future work achieved, practical problems of dyeing can be addressed. The problems include archiving and restoring traditional dyeing techniques, designing graphics of dyeing, education of the dyeing culture, etc. One of our important goal is that the real-world dyeing would be raised through these approaches. The dyeing frameworks in computer graphics should be used to enhance social interests in dyeing and enrich understanding but replace traditional real-world dyeing work. People can use the CG dyeing method for a wide range of aims. The methods include some factors of physics and design. In the future, visual simulation of dyeing would be able to develop as a theme between some scientic areas and be utilized widely.

7. References

Bruce Gooch, A. G. (2001). *Non-Photorealistic Rendering*, A K Peters Ltd.

Chu, N. S.-H. & Tai, C.-L. (2005). Moxi: real-time ink dispersion in absorbent paper, *ACM Trans. Graph.* 24(3): 504–511.

Chung, F. R. K. (1997). *Spectral Graph Theory*, American Mathematical Society. CBMS, Regional Conference Series in Mathematics, Number 92.

Curtis, C. J., Anderson, S. E., Seims, J. E., Fleischer, K. W. & Salesin, D. H. (1997). Computer-generated watercolor, *Computer Graphics* 31(Annual Conference Series): 421–430. URL: *citeseer.ist.psu.edu/article/curtis97computergenerated.htm*

Fick, A. (1855). On liquid diffusion, *Jour. Sci.* 10: 31–39.

Kunii, T. L., Nosovskij, G. V. & Vecherinin, V. L. (2001). Two-dimensional diffusion model for diffuse ink painting, *Int. J. of Shape Modeling* 7(1): 45–58.

Langmuir, I. (1916). The constitution and fundamental properties of solids and liquids. part i. solids., *Journal of the American Chemical Society* 38: 2221–2295.

Mitani, J. (2008). The folded shape restoration and the CG display of origami from the crease pattern, *13th international Conference on Geometry and Graphics*.

Morimoto, Y., Tanaka, M., Tsuruno, R. & Tomimatsu, K. (2007). Visualization of dyeing based on diffusion and adsorption theories, *Proc. of Pacific Graphics*, IEEE Computer Society, pp. 57–64.

Morimoto, Y., & Ono, K. (2010). Computer-Generated Tie-Dyeing Using a 3D Diffusion Graph, *In Advances in VisualComputing (6th International Symposium on Computer Vision - ISVC 2010 (Oral presentation)), Lecture Notes in Computer Science*, Springer, volume 6453, pages 707–718.

Polakoff, C. (1971). *The art of tie and dye in africa*, African Arts, African Studies Centre 4(3).

Press, W. H., Teukolsky, S. A., Vetterling, W. T. & Flannery, B. P. (1992). *Numerical recipes in C (2nd ed.): the art of scientific computing*, Cambridge University Press.

S.-H., B. (August 1997). Diffusion/adsorption behaviour of reactive dyes in cellulose, *Dyes and Pigments* 34: 321–340(20). URL: *http://www.ingentaconnect.com/content/els/01437208/1997/00000034/00000004/art00080*

Sakakibara, A. (1999). *Nihon Dento Shibori no Waza (Japanese Tie-dyeing Techniques)*, Shiko Sha (in Japanese).

Shamey, R., Zhao, X., Wardman, R. H. (2005). Numerical simulation of dyebath and the influence of dispersion factor on dye transport, *Proc. of the 37th conf. on Winter simulation, Winter Simulation Conference*, pp. 2395–2399.

Van den, B. R. (31 Deccember 1994). Human exposure to soil contamination: a qualitative and quantitative analysis towards proposals for human toxicological intervention values (partly revised edition), *RIVM Rapport* 725201011: 321–340(20). URL: *http://hdl.handle.net/10029/10459*

Vickerstaff, T. (1954). *The Physical Chemistry of Dyeing*, Oliver and Boyd, London.

Yoshiko, W. I. (2002). *Memory on Cloth:Shibori Now*, Kodansha International.

Wilson, B., Ma, K.-L. (2004). Rendering complexity in computer-generated pen-and-ink illustrations. *Proc. of Int. Symp. on NPAR*, ACM, pp. 129–137.

Wyvill, B., van Overveld, K. & Carpendale, S. (2004). Rendering cracks in batik, *NPAR '04: Proceedings of the 3rd international symposium on Non-photorealistic animation and rendering*, ACM Press, New York, NY, USA, pp. 61–149.

S. Xu, H. Tan, X. Jiao, F.C.M. Lau, & Y. Pan (2007). A Generic Pigment Model for Digital Painting, *Computer Graphics Forum (EG 2007)*, Vol. 26, pp. 609–618.

4

Lipid Role in Wool Dyeing

Meritxell Martí, José Luis Parra and Luisa Coderch
Institute of Advanced Chemistry of Catalonia (IQAC-CSIC), Barcelona
Spain

1. Introduction

The textile industry uses different fibers obtained from various animals, of which the wool from domesticated sheep *Ovis aries* is commercially the most important. Dyeing is one of the most important finishing procedures of wool processing. It almost invariably involves absorption of water-soluble colorants from aqueous solutions by the fibers. Diffusion is the process by which the colorant molecules penetrate the interior of the fibers.

Earlier workers studying the dye uptake of dyes by wool were mainly interested in the thermodynamics of the process, treating the wool fiber as a cylinder of uniform composition. Over the last decades there has been growing recognition of the importance of the diverse morphological structure of wool in determining its dyeing behavior (Rippon, 1992). Therefore a short wool structure description will be followed by a review of the role of this structure in wool dyeing.

The importance of the non-keratinous components of the fiber, specially the study of lipids present in the cell membrane complex prone us to emphasize in the lipid depleted wool, the modification of many properties and its behavior in the dyeing process. Besides the study of the mechanism of liposomes (made up with phospholipids) on wool dyeing could also help to elucidate the lipid role in wool dyeing.

1.1 Wool structure

Structurally, a wool fiber is an assembly of cuticle and cortical cells held together by the "cell membrane complex" (CMC). The surface of wool fibers consists of overlapping cuticle cells (Rippon, 1992). These are composed of two distinct major layers: the exocuticle and the endocuticle. The exocuticle consists of two sub-components, the A-layer, which is approximately 0.3 microns thick, and the B-layer, which is approximately 0.2 microns in thickness. These components differ mainly in the concentration of disulfide crosslinks (A-layer 35% half-cystine, B-layer 15% half-cystine and endocuticle 3% half-cystine).

Individual cuticle cells are surrounded by a thin membrane, the epicuticle, which is approximately 3–6 nm thick and accounts for around 0.1% of the total mass of the fiber. Although the epicuticle is proteinaceous, the surface of clean, untreated wool is hydrophobic. This property is the result of a thin layer of fatty acids (lipids) which are covalently bound to the surface of the epicuticle (the F-layer) (Naebe et al., 2010).

The cortex is made up of approx. 90% of keratin fibers, and is largely responsible for their mechanical behavior. It consists of closely packed overlapping cortical cells arranged parallel to the fiber axis. Cortical cells are approximately 100 μm long and 3–6 μm wide, and they are composed of rod-like elements of crystalline proteins (microfibrils) surrounded by a relatively amorphous matrix. The low-sulphur material, with a simple regular structure and without disulphide crosslinks, forms crystalline fibrils, which are embedded in a matrix of more complicated and crosslinked high-sulphur material. The fibrillar protein forms first, and the low-sulphur parts of the natural block copolymer crystallize in parallel rods separated by the high-sulphur domains. Then the rest of the high sulphur protein is formed and solidifies the matrix. Nature joins the two constituents of this natural composite in a special way. The low-sulphur protein molecules in the fibrils have high-sulphur domains that come out of the fibrils at intervals and are crosslinked to the rest of the amorphous matrix (Hearle, 1991).

As stated above, the cuticle and cortical cells are separated by a continuous network, the cell membrane complex. This accounts for approx. 3.5% of the fiber, is around 25 nm in width, and provides adhesion between the cells. The CMC has three major components: (i) an easily swollen "intercellular cement" (1.5%) of a crosslinked nonkeratinous protein (δ-layer); (ii) a lipid component (1%), which may be associated with β-layers; and (iii) a chemical resistant proteinaceous membrane (1%), which surrounds each cortical and cuticle cell.

1.2 Role of fiber structure in wool dyeing

This diversity of morphological structure is very important to determine the dyeing behavior of wool. Generally, when a textile substrate is dyed by an exhaustion method, the dyeing operation proceeds in three stages (Crank, 1956; Rippon, 1992):
1. diffusion of dye through the aqueous dyebath to the fibre surface
2. transfer of dye across the fibre surface
3. diffusion of dye from the surface throughout the whole fibre.

In order to obtain satisfactory shade development and fastness properties, complete penetration of dye into the fibre is essential. The rate at which this occurs is controlled by the rate of dye diffusion across the fibre surface and then throughout the whole interior. The rate can be markedly affected by altering the net charge on the fibre, by modifying the epicuticle or by altering the rate of diffusion within the fibre.

If the wool fibre is treated as a uniform cylinder, Fick's laws of diffusion (Crank, 1956) dictates that a plot of dye uptake versus the square root of time should be a straight line over most of the dyeing curve (Medley & Andrews, 1959). In the case of wool, however, the dyeing curve is initially concave and only becomes linear after some time. This observation led to the assumption that a "barrier", with a small capacity for dye, exists at the fiber surface (Medley & Andrews, 1959). The barrier was believed to be responsible for the non-Fickian dyeing isotherms obtained with wool (Medley & Andrews, 1959; Leeder, 1999).

Earlier workers identified the epicuticle with the barrier to dye penetration, thinking that this component constitutes a continuous membrane around the whole fibre (Lindberg et al., 1949; Milson & Turl, 1950). The barrier has also been ascribed to the whole cuticle (Makinson, 1968) and to the highly crosslinked A-layer of the exocuticle (Hampton & Ratte, 1979). All these suggestions regarding the nature of the barrier were based on a common belief that dyes must diffuse through the cuticle cells in order to reach the fiber cortex (i.e. the transcellular route shown in Figure 1. The epicuticle is not a continuous membrane, however, but surrounds each individual cuticle cell. Thus, gaps exist between the scales where the intercellular material extends to the exterior fiber surface (approximately 0.05% (Joko et al., 1985)).

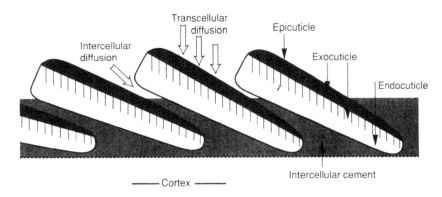

Fig. 1. Diffusion pathways for dyes into wool (Simmonds, 1955).

It has been suggested that lipids present at the cuticular junctions may hinder entry of dye into the fiber (Joko et al., 1985; Leeder et al., 1985a); for example, treatment of wool with potassium t-butoxide in anhydrous t-butanol, markedly improves the dyeing rate. This observation appears to be inconsistent with the fact that the anhydrous treatment is confined to the fiber surface, where it removes the F-layer from the epicuticle (Rippon & Leeder, 1986). The anhydrous alkali treatment would, however, be expected to remove lipids from the CMC at the point where this component extends to the fiber surface (Joko et al., 1985).

Different methods have been studied for comparative assessment of surface lipid removal from wool fabric; methanolic potassium hydroxide, anhydrous t-butoxide in t-butanol, and aqueous hydroxylamine (Negri et al., 1993; Ward et al., 1993; Dauvemann-Gotsche et al., 2000; Baba et al., 2001). Plasma treatments, which utilize a gaseous electrical discharge, are reported to be surface specific for wool fibers (Kan et al., 2004) and offer the potential of simple, clean, solvent-free, and inexpensive treatment (Thomas et al., 2005; Körner & Wortmann, 2005; Klausen et al., 1995; Kalkbrenner et al., 1990).

All these treatments (Thomas, 2007; Wakida et al., 1993, 1996, 1994; Lee et al., 2001; Höcker et al., 1994; Yoon et al., 1996; Jocic et al., 2005; Kan et al., 1998; Kan & Yuen, 2008; El-Zawahry et al., 2006; Kan, 2006) have shown to invariably increase the rate of uptake of dyes by wool. Some studies have revealed that little or no physical change to the surface structure of wool fibers results from treatment with plasma, whereas others have found significant damage (Thomas, 2007; Wakida et al., 1993; Yoon et al., 1996; Kan, 2006; Erra et al., 2002). The methods for producing plasmas vary considerably, and generally rely on the use of purpose-built equipment.

To obtain a better understanding of the uptake of dyes by plasma treated wool several aspects must be considered. A recent study has: (a) established conditions where only surface changes occur to wool fibers using a pilot-scale, commercial, atmospheric pressure plasma machine, and characterized those changes; and (b) examined the impact of the plasma treatment on the uptake of selected acid, 1:2 metal-complex and reactive dyes –

under adsorption conditions at 50 °C, as well as absorption conditions at 90 °C – with the aim of rationalizing the relationship among surface properties, dye structure and dye uptake (Naebe et al., 2010).

The dyes used were typical, sulfonated wool dyes with a range of hydrophobic characteristics, as determined by their partitioning behavior between water and n-butanol. Dye adsorption is a complex phenomenon involving both hydrophobic and electrostatic interactions. No significant effects of plasma on the rate of dye adsorption were observed with relatively hydrophobic dyes. In contrast, the relatively hydrophilic dyes were adsorbed more rapidly (and uniformly) by the plasma-treated fabric. It was concluded that adsorption of hydrophobic dyes on plasma-treated wool was influenced by hydrophobic interactions, whereas electrostatic effects predominated for dyes of more hydrophilic character. On heating the dyebath to 90 °C in order to achieve fiber penetration, no significant effect of the plasma treatment on the extent of uptake or levelness of a relatively hydrophilic dye was observed as equilibrium conditions were approached.

Extraction of normally scoured wool with lipid solvents also increases the dyeing rate (Medley & Andrews, 1959; Joko et al., 1985; Lindberg, 1953). This observation supports the concept that a lipid barrier to wool dyeing exists located at, or near, the fiber surface. A significant finding is that surface lipids appear to be concentrated mainly at the edges of the cuticle cells (Aicolina & Leaver, 1990).

Leeder, et al. (Leeder et al., 1985b) used specially synthetisized dyes to study the mechanism of wool dyeing. The metal-complex dyes contained platinum, palladium or uranium atoms, but in other respects were similar to conventional anionic wool dyes. The nuclear-dense, heavy metal atoms in the model dyes have a high electron-scattering power. This property enabled their location in the fiber to be determined with the transmission electron microscope at different stages of the dyeing process. This investigation provided the first unequivocal evidence that dye does, in fact, enter the wool fiber between cuticle cells, and also showed that dye diffuses along the nonkeratinous endocuticle and CMC early in the dyeing cycle.

The above finding supports the view that the cuticle (Makinson, 1968), probably the highly crosslinked A-layer of the exocuticle (Hampton & Rattee, 1979; Baumann & Setiawan, 1985), is a barrier to dye penetration, in that dyes are directed to the gaps between the scales in order to reach the cortex. It appears, however, that lipids present at the intercellular junctions are also a barrier to the diffusion of dyes into the nonkeratinous regions of the CMC (Leeder et al., 1985a).

After initial penetration into wool fibers, dyes must diffuse throughout the entire cross-section in order to obtain optimum colour yield and fastness properties. Several workers have suggested that the continuous network of the CMC provides a pathway for the diffusion of reagents into wool. Leeder and Rippon (Leeder & Rippon, 1982) have shown that the CMC swells in formic acid to a much greater extent than does the whole fiber. They suggested (Rippon & Leeder, 1986) that this disproportionately high swelling is the reason why dye is taken up very rapidly from concentrated formic acid.

The situation regarding the pathway for the dye diffusion into wool remained unresolved until the study by Leeder et al. (Leeder et al., 1985b) involving the transmission electron microscope, described above. This investigation demonstrated unequivocally the importance of the nonkeratinous components of the fiber in wool dyeing. After dye has entered the fiber between the cuticle cells, diffusion occurs throughout all the nonkeratinous regions of the CMC, the endocuticle and the intermacrofibrillar material. It is interesting that dye also appears in the nuclear remnants very early in the dyeing cycle, before dye can be seen in the surrounding cortical cells. The mechanism by which this occurs is not clear, but dyes may diffuse along "membrane pores". Kassenbeck (Zhan, 1980) has suggested that these pores connect the nuclear remnants with the endocuticle and the CMC.

The above findings on the importance of the nonkeratinous regions as pathways for diffusion of dyes into wool and other animal fibers have been confirmed by fluorescence microscopy (Leeder et al., 1990, Brady, 1985, 1990). However, the lower resolution of the light microscope, compared with the transmission electron microscope, restricts the amount of information that can be obtained by this technique.

Wortmann et al. (Wortmann et al., 1997) summarized different ideas about this subject. They justified that due to the CMC size in the fiber, intercellular pathway appears questionable. CMC makes up at most about 4-6% of the fiber structure, of which 1.5% are resistant membranes, 1.5% are lipids, and 1-3% are intercellular cement (Rippon, 1992). But only the two latter ones may be assumed to play a role in the intercellular pathway. After a revision of different experiments from several authors (Wortmann et al., 1997) and further references therein, they suggested that there is, in fact, little evidence for the intercellular pathway, which identifies intercellular diffusion as the primary pathway for dyes into fibers. Instead the results support their view that under normal dyeing conditions, diffusion proceeds through the nonkeratinous components of the fiber according to a restricted transcellular diffusion mechanism. With fibers that exhibit an intact epicuticle as a diffusion barrier, dyes will enter at the distal cuticle cell edges and will diffuse along the endocuticle and the CMC. Having reached the cortex, they then largely follow the intermacrofibrillar material and from there enter the nuclear remnants.

Swift (Swift, 1999) supports Wortmann et al. arguments with his microscope studies about precipitate silver sulphide in human hair by allowing aqueous silver nitrate to diffuse into fibers previously saturated under high pressure with gaseous hydrogen sulphide.

Rippon described the mechanism in his review (Rippon, 1992): after entering the fiber between cuticle cells, dye diffuses throughout all the nonkeratinous regions, including the CMC, endocuticle, and intermacrofibrillar material. Dye also appears in the nuclear remnants early in the dyeing cycle. Dye molecules progressively transfer from the nonkeratinous regions into the sulphur-rich proteins of the matrix surrounding the microfibrils within each cortical cell, and also from the endocuticle into the exocuticle. At equilibrium, the nonkeratinous proteins, which were involved in the early stages of dyeing, are virtually devoid of dye. Rippon also enhanced that the intercellular cement component of the CMC, the only continuous phase in wool, provides a pathway for dyes to reach the cortical cell located inside the fiber, this occur while dye is simultaneously diffusing along the other nonkeratinous regions of the fiber, and indeed, while dye is also diffusing to its equilibrium location within the high-sulphur proteins of the matrix (Rippon, 1999).

2. Results and discussion

2.1 Wool modification due to internal lipid extraction

A number of studies have been performed in our laboratories based on the extraction, analysis and structure of the internal lipids and isolated ceramides from wool (Ramírezet al.,

2008a, 2008b; Coderch et al., 2002, 2003; Méndez et al., 2007; Fonollosa et al., 2004). Internal wool lipids have been shown to form stable liposomes (Fonollosa et al., 2000; Körner et al., 1995; Ramírez et al., 2009a) and are supposed to be arranged in the wool fiber as lipid bilayers.

Raw Spanish Merino wool was Soxhlet-extracted with chloroform/methanol azeotrope (Martí et al., 2010), in order to obtain wool mostly depleted of internal lipids. The lipids extracted were quantitatively analyzed by TLC-FID so that the main lipid families were separated and quantified.

It was observed that the percentage of lipids analyzed was 0.96% o.w.f., and the main compounds are free fatty acids 19.68%, sterols 6.45% and polar lipids where the ceramides 65.03% are included. It is important to bear in mind that the total internal lipids account for 1.5% of total fiber weight (Rivett, 1991) and the total extracted internal lipids only account for 0.96% of the total fiber.

Analytical and physicochemical studies reveal considerable resemblance between the internal wool lipids (IWL) and the lipids from the stratum corneum of the human skin (Coderch et al., 2003; Schaefer & Redelmeier, 1996; Schürer et al., 1991; Kerscher et al., 1991). IWL are present in about 1.5% of fiber weight and are rich in ceramides, cholesterol, free fatty acids and cholesteryl sulfate (Hearle, 1991). The intercellular lipids of the stratum corneum play a vital role in the barrier function of human skin by protecting it from the penetration of external agents, as well as by controlling the transepidermal water loss, which maintains the physiological skin water content (Kerscher et al., 1991; Rivett, 1991; Elias, 1981). In order to obtain IWL extracts with a large amount of ceramides, different extraction methodologies such as Soxhlet with diverse organic solvent mixtures or supercritical fluid extraction with CO_2 and several polarity modifiers have been optimized at laboratory and pilot plant levels (Ramírez et al., 2008a, 2008b; Coderch et al., 2002; Petersen, 1992).

Besides the chemical analyses of wool extracts, chemical and mechanical evaluations of extracted wool have been carried out. Residual grease, whiteness index, fiber diameter, fiber length, cleaning tests, alkaline solubility, bundle tenacity and drafting forces, abrasion resistance, pilling tests, and pore size have been determined. Few significant changes have been obtained in most of the assays between non-extracted and solvent extracted fibers. However, the higher abrasion resistance of extracted fabrics, the longer fiber length and the lower alkaline solubility of most lipid extracted wools should be noted (Ramírez et al., 2008b, 2009b; Petersen, 1992).

Additional analyses of the extracted fibers have been performed. Parameters such as yield, fibril and matrix viscoelastic behavior, deformation work and breaking elongation have highlighted the effect of IWL on the fiber mechanical properties. The IWL extraction has increased yield tenacity and decreased the elongation at break of the fibers, maintaining the feasibility of extracted wool for textile purposes (Martí et al., 2007).

Changes in hydrophobicity in IWL extracted fibers could be important in the dyeing process. Therefore, hydrophobicity of extracted and untreated wool fabrics was assessed by the wetting time test in order to ascertain whether the epicuticular lipids were removed during chloroform/methanol Soxhlet extraction. The results obtained were the same in two samples. The extracted and the non-treated wool fabrics remained on the water surface for more than 48 hours. This finding indicates that the epicuticular lipid layer is intact despite Soxhlet IWL extraction (Martí et al., 2010).

Accordingly, a contact angle of wool fibers was also performed. Five non-treated and five extracted fibers were analyzed and the mean values of perimeter and contact angle evaluated. The contact angles of the two wool fibers are similar, $87.51°$ (±3.1) for the non-treated and $86.94°$ (±3.8) for the extracted sample. Despite a non-significant reduction in hydrophobicity of extracted wool fiber, it may be concluded that the superficial hydrophobicity of extracted wools was not modified under these experimental conditions (Martí et al., 2010).

Moreover, in the absence of IWL, the extracted wool fibers absorbed more water. This behaviour was demonstrated by TG (Thermogravimetry) study and the DSC analysis (Differential Scanning Calorimetry), where the amount of water increases when the extraction time is longer (Martí et al., 2007). DSC is a method commonly used to determine crystallinity in polymers and involves measuring the melting enthalpy. Wool fibres have been studied by this technique in several papers (Cao et al., 1997; Spei & Holzem, 1987; Wortmann & Deutz, 1993, 1998; Wortmann, 2005). There are currently difficulties in accurately measuring the melting transition of wool owing to the fibrous nature of the sample, (for exemple the level of cystine content could have an influence on DSC analysis (Rivett, 1991), and to the moisture sensibility of the thermal transitions (Haly & Snaith, 1967). The differences in DSC parameters can be related to differences in the matrix material that is the non-helical parts of the intermediate filaments (IFs), the material existing between the IFs, and all other amorphous, morphological components (Wortmann, 2005).

For Merino wool the endotherm is often bimodal. Wortmann et al. confirmed that ortho-cortical cells have a lower melting point than para-cortical cells (Wortmann & Deutz, 1998), which could account for the bimodal peak. (Cao et al., 1997) have investigated the origin of this bimodal endotherm and have presented an alternative interpretation, in which the bimodal peak arises from the overlapping of the melting endotherm of α-form crystallites with the thermal degradation of other histological components.

A research was focused on the changes in the wool structure when IWL were extracted in order to better understand the bimodal endotherm peak behaviour when other histological compounds were extracted from wool. The decreased ΔH_D could mean that methanol extracts part of the amorphous α-form keratin, whereas the high temperature could mean that the rest of crystalline material was more stable than that of the non-extracted wool. This phenomenon could be related to the high abrasion resistance obtained for these samples (Martí et al., 2005). In this work, weakly marked endotherms were obtained in all the samples studied with small enthalpy differences. Therefore, it seems that our results may lend support to the melting endotherm corresponding to the differential melting behavior of the α-form crystallines in the domains of ortho- and para-cortical cells as affirmed by Wortmann et al. (Wortmann & Deutz, 1993, 1998) and by the results of Manich et al. (Manich et al., 2005).

An additional part of our study was focused on the determination of pore size of treated wools in order to analyse the possible pore modification due to lipid extraction. The technique used was thermoporometry, which is based on the determination of the melting temperature of imbibed water for different pore sizes. The cumulated pore volume of the extracted wool fibers increased, indicating that extraction of material from the CMC occurred. From all these results it seems that the morphological modification of extraction would exert an influence on the dyeing of wool lipid extracted fibers.

2.2 Dyeing process of lipid depleted wool fibers

As it was previously mentioned in section 1.3, a number of theories have been proposed concerning the influence of different wool compounds on the dyeing process. According to Wortmann (Wortmann et al., 1997) and Swift (Swift, 1999) the main pathway for dye diffusion is through the nonkeratinous components (endocuticle, intermacrofibrillar matrix and nuclear remnant zones) according to a restricted transcellular diffusion mechanism. However, for Rippon (Rippon, 1999) and Leeder (Leeder, 1999), the intercellular cement is the pathway by which dye molecules reach cortical cells. Therefore, a study of the influence of lipid extraction on the dyeing behavior of wool could help to lend support to one of these two theories.

Extracted and untreated raw wools were conventionally dyed with two acid dyes in order to elucidate the interaction of the dyes and the chemical wool structure with and with less internal wool lipids (0.96% of lipids extracted (Martí et al., 2010)).

The kinetics of extracted and non-extracted wool dyed at 98°C under conventional conditions by two dyes Acid Green 25 (hydrophilic) and Acid Green 27 (less hydrophilic) (Martí et al., 2004) were compared (Figure 2).

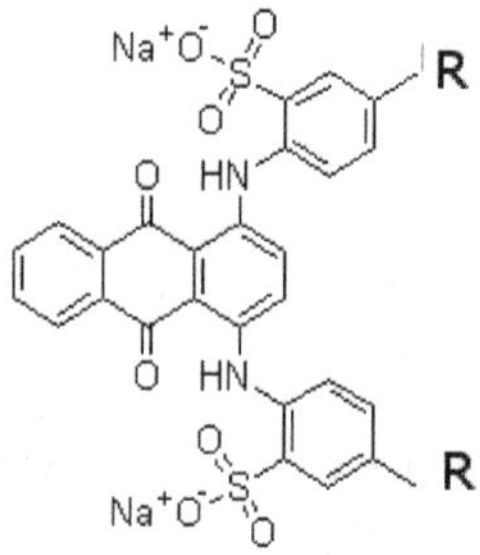

Fig. 2. Molecular structure of the acid dyes used: C.I. Acid Green 27 (R=C_4H_9) and C.I. Acid Green 25 (R= CH_3)

Partition coefficient measurements of the dyes were previously evaluated (Martí et al., 2004) and results (4.2 for Acid Green 25; 1.3 for Acid Green 27) indicated a large difference between these two dyes in their affinities to the polar or non-polar environment. Some bath aliquots were analyzed as the temperature increase. Figure 3a shows the dye exhaustion of Acid Green 25 in untreated and extracted wool. It can be seen that the more hydrophilic dye has high dye exhaustion (about 99%), at 70°C, and the untreated wool attained the highest dye exhaustion at 98°C.

Acid Green 27 (Figure 3b), the big molecular structure dye (less hydrophilic), had a different behavior when the temperature increases; the extracted wool had lower dye exhaustion than the untreated wool during the dyeing process. The diminution of lipids could account for the faster penetration of the small molecular structure dye into the extracted wool fibers than into untreated wool (which retains its original composition). The interaction of this dye with short carbon chains (–CH_3) and its penetration into the fibers without IWL were increased, because the lipids may act as a barrier for more hydrophilic. This contrasts with the big molecular structure dye (–C_4H_9) which had lower affinity with the modified wool fiber, because the hydrophobic forces were absent, resulting in a decreased dye exhaustion.

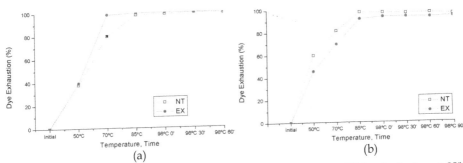

(a)   (b)

Fig. 3. (a) Acid Green 25 kinetic dyeing at 98°C and (b) Acid Green 27 kinetic dyeing at 98°C.

In order to corroborate these results, another kinetic study was performed in which a final temperature of 70°C was attained, given that a maximum exhaustion was obtained for Acid Green 25 at this temperature. Therefore, two dye processes at 70°C were performed with the same dyes (Figures 4). The dyeing conditions were the same as the earlier kinetic dyeing except for the final temperature of 70°C.

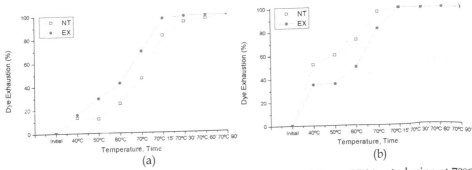

(a)   (b)

Fig. 4. (a) Acid Green 25 kinetic dyeing at 70°C and (b) Acid Green 27 kinetic dyeing at 70°C.

Again, the extracted fiber was dyed faster than the untreated fiber with Acid Green 25, yielding 92% of dye exhaustion after 15 minutes at 70°C (Figure 4a). These results confirmed the faster penetration of the hydrophilic dye when the fiber is mostly depleted of lipids and becomes more hydrophilic.

In Figure 4b, the dye behavior contrasts with that shown in Figure 4a. The less hydrophilic dye, the faster the penetration into the fiber in the untreated sample, which maintains its lipid internal structure and has a hydrophobic character.

In order to assess dye penetration mechanism of the lipid depleted wool fibers, a microscopic study of the Acid Green 25 and Acid Green 27 dye diffusion was performed (Figure 5). Similar staining behavior was obtained for the two dyes when applied to lipid-extracted wool at the end of the dyeing processes. High penetration was observed with no ring dyeing, which supports a correct penetration for the two dyes on the lipid-extracted wool fibers.

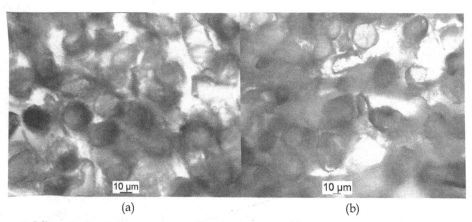

(a)   (b)

Fig. 5. Micrographs of cross sections of Acid Green 25 (a) and Acid Green 27 (b) lipid extracted wool dyed samples.

K/S values of final dyed goods were also evaluated for the fibers with and with less internal lipids at the two final dyeing temperatures of 98°C and 70°C with the two dyes. Results indicate the same tendency as the dye exhaustion values; however, the dyeing differences are much marked. While the extracted fibers present a lighter color when dyed with the less hydrophilic dye Acid Green 27, the opposite occurs with the more hydrophilic one, Acid Green 25. This higher affinity for the hydrophilic dye to the extracted fiber is even maintained at 70°C, the final dyeing temperature.

To evaluate the visual appearance of the final dyed flock samples, the CIELAB L* a* b* values were obtained. Hue becomes more blue-green in this case for the two dyestuffs. In the case of Acid Green 27, the influence of the extraction is more marked in the final color, between non-extracted and extracted samples, than with the Acid Green 25.

Color fastness was also followed to evaluate dye fastness of wool fabrics with or with less internal lipids. Slightly lower color-fastness values were obtained for the dyed fibers with less internal lipids in the staining degree onto diacetate, polyamide and polyester. In addition, the fastness values for less hydrophilic dye Acid Green 27 are in general slightly lower than for Acid Green 25; this might be due to low interaction between more hydrophilic extracted fiber and the less hydrophilic dye (Acid Green 27). In this case, when the dyed fabric undergoes washing conditions the dye is easily released from the lipid-free fibers. Slight differences in the staining degrees are observed between Acid Green 25 dyed fabrics.

According with the two theories proposed concerning the influence of different wool compounds on the dyeing process, the null hydrophobic surface modification due to lipid extraction and the dyeing results obtained lend support to the intercellular cement as the main dye pathway. The internal lipids, which account for only 33% of the continuous phase of wool and about 1.5% o.w.f., play a major role in the dyeing mechanism.

The results obtained show the different dye behaviors of wool fibers with and almost without IWL. Two dyestuffs were used, the only difference being in the length of their chains; a shorter one that was more hydrophilic (Acid Green 25) and a longer one that was more hydrophobic (Acid Green 27). Maximum dye exhaustion, (about 98%), was achieved at

70°C in extracted wool when a hydrophilic dye was used, with 100% of final exhaustion and a small difference in the hue (CIELAB values). However, slightly lower dye exhaustion values (93% *versus* 96%) were obtained at temperatures over 85°C in the dyeing process in extracted wool when a hydrophobic dye was applied. These different dyeing behaviors may be attributed to the interaction between IWL and the dyestuff.

It may be deduced that the depletion of the hydrophobic internal lipid structure in wool leads to a more hydrophilic pathway. Therefore, the hydrophilic dye easily penetrates into the extracted fiber in contrast to the dye with longer alkyl chains. Since contact angle and wetting time measurements show no modification in the hydrophobicity character of the surface, these results support the theory that the intercellular cement is the main pathway for dyes, highlighting the role of the IWL in this process.

These findings indicate a similar dyeing behavior of wool fibers mostly depleted of internal lipids when a hydrophobic dye was used, and a marked increase in dye exhaustion when a hydrophilic dye was applied. This strategy proves useful in reducing the final dyeing temperature and in mitigating the fiber damage without impairing the washing fastness of the fibers. However, consideration should be given to the color differences obtained between the dyed samples at different temperatures.

In the same way, Telegin et al. (Zarubina et al., 2000) have indicated that hydrophilic/lipophilic properties of acid dyes predetermine the mechanism of their interaction with wool fibers. Temperature changes in sorption of hydrophilic dye and high values of affinity of lipophilic dye support the idea that nonkeratinous components control the transport of acid dyes into the fiber.

2.3 Wool dyeing with liposomes

In our laboratories the role of wool lipid on wool dyeing was already detected on the dyeing process using phosphatidylcholine liposomes as a dyeing auxiliary (Martí et al., 2004). The same two acid dyes of the work presented in the previous section with the only difference being in the length of their chains; the shorter one being more hydrophilic (Acid Green 25) and the longer one being more hydrophobic (Acid Green 27) were used to study the mechanism of wool dyeing.

Liposomes are vesicular colloidal particles with inner water volume separated from bulk water with self-closed lipid bilayers. Having both hydrophilic and hydrophobic compartments in their structure, liposomes can loaded with substances of different polarity ranging from water-soluble molecules, which are entrapped in the inner aqueous space of liposomes, to hydrophobic molecules which can be dissolved in non-polar bilayer interior. Due to their effective encapsulation capacity liposomes have found numerous applications in various fields, as drug delivery vehicles (Lasic, 1993). In recent years, liposomes have been used in textile industry as dyeing auxiliaries, mainly for wool dyeing (Martí et al., 2001; Montazer et al., 2006; Barani & Montazer, 2008). Dyeing of wool and wool blends along with liposomes has demonstrated better quality, energy saving, and lower environmental impacts. The temperature of dyeing of pure wool and wool blends could be reduced and there was less fiber damage. Moreover, dyebath exhaustion was shown to be greater than 90% at the low temperature (80°C) used resulting in significant saving in energy costs. The impact of the dyeing process on the environment was also much reduced with chemical oxygen demand (COD) being reduced by about 1000 units (Rocha Gomes et al., 1997; Coderch et al., 1999a, 1999b; Martí et al., 1998, 2001; de la Maza et al., 1998; Montazer et al., 2007).

The self-assembling behavior of liposomes and their physicochemical stability at acidic pH values (4.0-5.0) and temperature range (40-90 °C) demonstrated that the liposomes are stable under experimental conditions of the dyeing process. The temperatures used in the dyeing process are always higher than the transition temperature of lipids forming liposomes ($\sim$-10°C). This implies the continuous fluid state of these lipids maintaining the vesicles without structural modifications.

Transition temperature of the internal lipids was previously evaluated ($\sim$ 45°C) (Méndez et al., 2007); this is important to modulate dye diffusion because they are the main components of the CMC (Cell Membrane Complex) which is presumably the dye pathway.

Liposomes can influence the dyeing process through their interactions with the wool fibers and at the same time with the dyestuffs. To elucidate the effects of liposomes on each of these substrates, the dyeing kinetics for the dyes Acid Green 25 and Acid Green 27 were compared in three experimental protocols: (i) in the presence of phosphatidylcholine (PC) liposomes, (ii) without liposomes, and (iii) with wool previously treated with PC liposomes.

For untreated wool fibers, a retarding effect of liposomes at the first stages of the dyeing process was observed in the case of the hydrophobic dye Acid Green 27 when the liposomes were present in the bath. At the end of the process, the same (Acid Green 27) or higher (Acid Green 25) dye exhaustion values were obtained when compared with the dyeing process without liposomes in the bath (Figure 6). At the first stages of the dyeing process, the higher retarding effect of the liposomes with the Acid Green 27 could be due to the higher affinity of the hydrophobic dye to the liposomes present in the bath in comparison with the wool fiber. In fact, the previous studies on the liposome-dye interaction and its influence on dyeing kinetics demonstrated a retarding effect on the dye exhaustion due to dye accumulation in liposomes, which takes place in measurable amounts even in the presence of wool (Simonova et al., 2000).

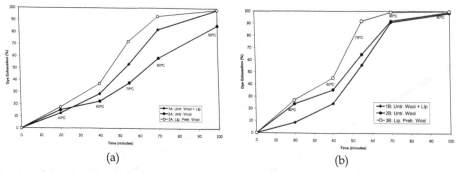

(a)   (b)

Fig. 6. Exhaustion kinetics of Acid Green 25 (a) and Acid Green 27 (b). 1: Untreated wool in the presence of liposomes, 2: Untreated wool, 3: Liposome pretreated wool.

However, the most striking feature is an increase in the dye exhaustion for the two dyes at all stages of the dyeing process when wool was previously treated with liposomes. The liposome-wool interaction responsible for this behavior could be explained by possible structural changes in the CMC of the fiber due to the previous PC absorption and/or the eventual wool lipid solubilization, which could increase wool permeability for the dye molecules.

Therefore, different experiments were performed to elucidate the liposome–wool interaction in the wool dyeing process. Liposome absorption by the wool fibers was followed by quantifying the amount of the total phosphorus in the bath at different stages of the liposome pretreatment. In addition, DSC measurements were carried out using dimyristoylphosphatidylcholine (DMPC) liposomes as a probe to monitor changes in their thermotropic behavior that may be related to the liposome–wool interaction. Structural changes in the CMC of wool fiber were also evaluated by analyzing the lipids extracted from the liposome-treated wool fibers in order to determine whether PC is actually absorbed by the wool fibers and whether the composition of the internal wool lipids is modified.

Liposome absorption by the wool fibers was determined by the quantitative total phosphorus analysis in aliquots taken from the bath under conditions of the liposome pretreatment. A very quick PC absorption (24%) in the first stage of the process was observed, being also especially important in the last 30 min of incubation at 90 °C to achieve a 39% PC absorption.

The influence of wool on the thermotropic properties of liposomes was studied using the DMPC liposomes as a DSC probe. Heating of the DMPC liposomes (1% oww) with wool at 70-90 °C resulted in complete disappearance of the DMPC signal from the DSC thermograms.

A deep DSC study showed that some liposome-active material is actually solubilized from wool, even at low temperatures, the releasing process being more efficient in the presence of liposomes. A strong effect of this solubilized material on the phase behavior of liposomes implies that this substance has a high affinity to lipid bilayers and thus may originate from the lipid constituents of the cell membrane complex of the wool fibers.

In fact, the experiments performed with liposomes prepared from IWL have clearly shown that these lipids also exert a strong broadening effect on the DMPC thermogram (Figure 7). This supports our supposition that wool being incubated with liposomes releases into the incubation bath some lipid material (presumably polar lipids) that enters the liposome membrane and affects drastically its properties.

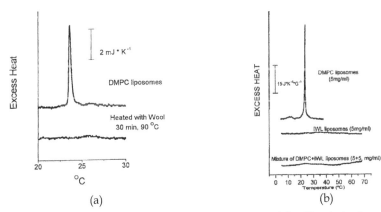

(a) (b)

Fig. 7. (a) DSC thermograms of DMPC liposomes heated at 90°C for 30 min in the presence or absence of wool, (b) DSC thermograms of DMPC liposomes, IWL liposomes, and a mixture of DMPC and IWL liposomes.

A release of the lipid material from wool should be accompanied by changes in the lipid composition of wool fibers. The results of lipid extracted from different treated wool performed by TLC/FID (Martí et al., 2004), confirmed that the decrease in the liposome content of the incubation bath observed on pretreatment of wool with liposomes was actually accompanied by the PC absorption by the wool fibers. Being this PC strongly bound to the wool fibers, seemingly due to its incorporation into the lipid domains of the CMC.

The amounts of free fatty acids and sterols extracted from all wool samples were very similar regardless of the experimental dyeing conditions used and the presence of liposomes in the bath. However, a substantial decrease in polar lipids was detected even when wool was subjected to the dyeing conditions in the absence and in the presence of PC liposomes in the incubation bath. These experiments also showed that the removal of polar lipids from wool was accompanied by simultaneous penetration of PC into the wool fibers when incubation was carried out in the presence of liposomes. Since the polar lipids consisting mainly of ceramides and cholesterol sulfate significantly differ from PC in chemical structure and membrane behavior, the authors supposed that such a substitution should greatly affect those properties of the CMC that govern the permeability of wool to dye molecules.

Indeed, it had been shown by electron paramagnetic resonance (EPR) measurements on mixed IWL/PC liposomes that the presence of PC, especially at low amounts (10 wt %), greatly fluidized the lipid bilayer at any temperature and decreased the enthalpy of the main phase transition of IWL from an ordered gel state to a liquid-crystalline fluid state (Fonollosa et al., 2000). If the dye, owing to its amphiphilic nature and some affinity to the lipid bilayer, was able to diffuse along the CMC through lipid domains, then an agent that increases their fluidity would facilitate the dye penetration deep into the wool fiber.

This was the effect that had been observed for the liposome pretreated wool (see Figure 6), which contains the highest amount of PC absorbed by the wool fibers. However, when the dyeing was performed in the presence of PC liposomes two different processes seemed to compete against each other. On one hand, PC tent to enter the wool fibers, and on the other hand, dye had affinity to the liposomes, the latter process being more pronounced for the more hydrophobic dye. Therefore, at initial stages of the dyeing process, when the amount of PC incorporated into wool was low, the retarding effect of liposomes on the dye exhaustion kinetics predominates, which was especially obvious in the case of the hydrophobic dye Acid Green 27.

Our findings indicated that the presence of liposomes in the dyeing bath promotes retention of the two dyes investigated at low temperature, this effect being more important in the case of the hydrophobic dye. It appears that the hydrophobicity of liposomes competes with that of the wool fibers so that the more hydrophobic dye was retained in the dyeing bath to a greater extent. This study had also shown that liposomes and wool interacted actively with each other mainly at high temperature (above the internal lipid transition temperature ~45°C) with wool polar lipids / PC interchange resulting to a high fluid lipid bilayer in the CMC. This interaction resulted in such a modification both of liposomes and wool fibers that eventually favors the dyeing process, with a dye retardant effect at the beginning of the process and an increase of final dye exhaustion (Figure 8).

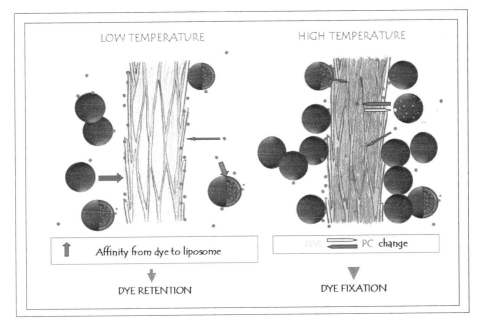

Fig. 8. Liposome−wool-dye interaction.

The detailed study of the liposome−wool interaction with the physicochemical methods revealed that an exchange of some lipid material between liposomes and wool fibers might occur (Figure 8). It was demonstrated that phosphatidylcholine from liposomes was absorbed by wool when wool fibers were subjected to the liposome treatment. On the other hand, a membrane-active factor was released from wool into the water phase, the release being highly intensified in the presence of liposomes. The strong effect exerted on the phase behavior of liposomes implied that this material has a high affinity to lipid bilayers and may originate from the lipid constituents of the cell membrane complex of the wool fibers. This assumption was confirmed by model experiments with liposomes prepared from internal wool lipids. As far as the cell membrane complex plays a key role in penetration and diffusion of dyes into the wool fibers, these results may be helpful in a better understanding of lipid pathways of wool dyeing (Martí et al., 2004).

3. Conclusion

Most plasma treatments invariable increase the rate uptake of dyes by wool. They do not only remove the covalently bond fatty layer (F-layer) which end up to a less hydrophobic wool surface, but also results in exposure of the underlying hydrophilic protein material, which increase the effectiveness of the ionic interaction between protein of the epicuticle and exocuticle and the more hydrophilic molecules. Modification of external linked lipids from the epicuticle showed the importance of both hydrophobic and electrostatic interactions in the dye absorption (Naebe et al., 2010). Hydrophobic dye had little impact of the plasma treatment on dye uptake. It appeared that, for the plasma-treated wool, there was still a sufficient number of hydrophobic groups on the exposed surface of the epicuticle to facilitate this mechanism of dye adsorption. For the more polar disulfonated and trisulfonated dyes, it appeared that electrostatic effects were more important for adsorption than were hydrophobic effects.

With respect to modification of internal wool lipids, the fundamental role of the internal lipids in the penetration of dyestuffs into fibers should also be noted. A very different dye behavior was obtained for wool fibers with and without internal lipids. Maximum dye exhaustion was achieved in extracted wool when a hydrophilic dye was used. However, lower dye exhaustion was obtained in the dyeing process of extracted wool when a hydrophobic dye was applied. These different dyeing behaviors may be attributed to the interaction between the internal wool lipids and the dyestuff. It was deduced that the depletion of the hydrophobic internal lipid structure in wool leads to a more hydrophilic pathway. Therefore, hydrophilic dye may easily penetrates into the extracted fiber in contrast to the dye with longer alkyl chains.

The presence of liposomes in the dyeing bath promoted retention of the two dyes investigated, this effect being more important in case of the big molecular structure dye. It appears that the hydrophobicity of liposomes competes with that of the wool fibers so that the bigger molecular structure dye was retained in the dyeing bath to a greater extent. This study (Martí et al., 2004) also showed that liposomes and wool interact actively to each other. This interaction resulted in such a modification both of liposomes and wool fibers that eventually favors the dyeing process.

According to the arguments above exposed and taking into account the complexity of the fiber structure, it can be concluded that there are several barriers for dyeing penetration being important the F-layer of the epicuticle for the diffusion of dye through the aqueous dye bath to the surface. However, the results obtained from the study of the dye process of lipid depleted wool fibers and the liposome assisted dye uptake support the theory that the intercellular cement is the main pathway for dyes, highlighting the role of the internal wool lipids in this process mainly in the transfer of dye across the fibers.

4. Acknowledgment

The authors are indebted to Ms. I. Yuste for technical support. And part of this work was supported by funds from the INTAS Project 97-0487 and by Spanish National Project (Ministerio de Educación y Ciencia) CTQ-PPQ2009-13967-C03-01.

5. References

Aicolina, L.E. & Leaver, I.H. (1990). *Proceedings of 8th Int. Wool Text. Res. Conf.,* Vol. 4, pp. 297, Christchurch, NZ, February 1990.

Baba, T.; Nagasawa, N.; Ito, H.; Yaida, O. & Miyamoto, T. (2001). Changes in the Covalently Bound Surface Lipid Layer of Damaged Wool Fibers and their Effects on Surface Properties, *Textile Res J.,* Vol. 71, No. 4, pp. 308-312, ISSN 1746-7748.

Barani, H. & Montazer, M. (2008). A review on applications of liposomes in textile processing. *J. Liposome Res.* Vol 18, No.3, pp. 249-162, ISSN 0898-2104.

Baumann, H. & Setiawan. (1985). Topochemistry of Dyeing and Chemical Processing of Wool. *Proceedings 7th Int. Wool Text. Res. Conf.,* Tokyo, Vol.5, pp.1-13.

Brady, P.R. (1985). Dyeing Wool at Low Temperature for Minimum Damage. *Proceedings of 7th Int. Wool Text. Res. Conf.*, Tokyo, Vol. 5, pp. 171-180.

Brady, P.R. (1990). Penetration Pathways of Dyes into Natural Protein Fibres. *Proceedings of 8th Int. Wool Text. Res. Conf.*, Christchurch, NZ, February 1990, vol. 4, pp. 217-226.

Cao, J.; Joko, K. & Cook, J.R. (1997). *Textile Res. J.* Vol.67, pp. 117-, ISSN 1746-7748

Coderch, L.; Manich, A.M.; Martí, M.; de la Maza, A.; Parra, J.L. & Serra, S. (1999a). Complemetary study of optimizing a wool dyeing process with commercially available liposomes. *Textile Res J.*, Vol. 69, No. 10, pp. 789-790, ISSN 1746-7748.

Coderch, L.; Martí, M.; de la Maza, A.; Manich, A.M.; Parra, J.L. & Serra, S. (1999b). Potencial of liposomes to aid the quest for cleaner dyeing. *Wool Record*. Dec: 26.

Coderch, L.; Fonollosa, J.; Martí, M.; Garde, F.; de la Maza, A. & Parra, J.L. (2002). Extraction and analysis of ceramides from internal wool lipids. *J. Am. Oil Chem. Soc.* Vol. 79, pp. 1215-1220, ISSN 0003-021X.

Coderch, L.; Bondia, I.; Fonollosa, J.; Méndez, S.; and Parra, J.L. (2003). Ceramides from Wool: Analysis and structure, *IFSCC Mag.*, 6, 117-123, ISSN 1520-4561

Crank, J. (1956) *The mathematics of diffusion* (Oxford: Clarendon Press)

De la Maza, A.; Coderch, L.; Manich, A.M.; Martí, M.; Parra, J.L. & Serra, S. (1998). Optimizing a Wool Dyeing Process with an Azoic 1:2 Metal Complex Dye Using Commercial Available Liposomes. *Textile Res. J.* Vol. 68, No. 9, pp. 635-642, ISSN 1746-7748.

Dauvermann-Gotsche, C.; Evans, D.J.; Corino, J.L. & Korner, A. (2000). Labelling of 18-Methyleicosanoic Acid Containing Proteo-lipids of Wool with Monomaleimido Nanogold. *Procedings 10th Int. Wool Text. Res. Conf.* Aachen, German Wool Research Institute, Vol. St-10, pp. 1-10, ISBN 3-00-007905-X.

Elias, P.M. (1981). Lipids and the Epidermal Permeability Barrier. *Arch. Dermatol. Res.* Vol. 270, pp. 95-117, ISSN 0340-3696.

El-Zawahry, M. M.; Ibrahim, N. A. & Eid, M. A. (2006). The Impact of Nitrogen Plasma Treatment upon the Physicochemical and Dyeing Properties of Wool Fabric, *Polym. Plast. Technol. Eng.* Vol. 45, No. 10, pp. 1123–1132, ISSN 0360-2559.

Erra, P.; Jovancic, P.; Molina, R.; Jocic, D. & Julia, M. R. (2002). Study of Surface Modification of Wool Fibres by Means of SEM, AFM and Light Microscopy, in "Science, Technology and Education of Microscopy: an Overview," Vol. II, pp. 549–556, Formatex: Badajoz, Spain.

Fonollosa, J.; Campos, L.; Martí, M.; de la Maza, A.; Parra, J.L. & Coderch, L. (2004). X-ray diffraction analysis of internal wool lipids. *Chem. Phys. Lipids*, Vol. 130, 159-166, ISSN 0009-3084.

Fonollosa, J.; Martí, M.; de la Maza, A.; Sabés, M.; Parra, J.L. & Coderch, L. (2000). Thermodynamic and Structural Aspects of Internal Wool Lipids. *Langmuir*, Vol. 16, No. 11, pp. 4808-4812, ISSN 0743-7463.

Haly A.H.; Snaith, J.W. (1967). Differential Thermal Analysis of Wool –The Phase-Transition Endotherm Under Various Conditions 1. *Textile Res. J.* Vol. 37, No. 10, pp. 898-907, ISSN 1746-7748.

Hampton, G.M. & Rattee, I.D. (1979). Surface Barrier Effects in Wool Dyeing Part I –The Location of the Surface Barrier. *J. Soc. Dyers Colour.* Vol.95, No. 11, pp. 396-399, ISSN 0037-9859.

Hearle, J.W.S (1991). Understanding and control of textile fiber. *J. Appl. Polym. Symp.*; ISSN 0271-9460.

Höcker, H.; Thomas, H.; Küsters, A., and Herrling, J. & Färben von Plasmabehandelter Wolle. (1994). Dyeing of plasma treated wool. *Melliand Textilber.* Vol. 75, pp. 506-512.

Jocic, D.; Vilchez, S.; Topalovic, T.; Molina, R.; Navarro, A.; Jovancic, P.; Julia, M. R. & Erra, P. (2005). Effect of Low-temperature Plasma and Chitosan Treatment on Wool Dyeing with Acid Red 27. *J. Appl. Polym. Sci.* Vol. 97, pp. 2204-2214, ISSN 1097-4628.

Joko, K.; Koga, J. & Nuroki, N. (1985). The Interaction of Dyes with Wool Keratin – The Effect of Solvent Treatment of Dyeing Behavior. *Proceedings 7th Int. Wool Text. Res. Conf.*, Tokyo, Vol.5, pp.23-32.

Kalkbrenner, U.; Körner, A.; Höcker, H. & Rivett, D.E. (1990). Studies on the Lipid Composition of the Wool Cuticle. *Proceedings 8th Int. Wool Text. Res. Conf.* Chistchurch, Wool Research Organization of New Zealand, Vol. 1, pp. 398-407.

Kan, C.W.; Chan, K. & Yuen, C.W.M. (2004). Surface Characterization of Low Temperature Plasma Treated Wool Fiber – The Effect of the Nature of Gas. *Fibers Polymers.* Vol. 5, No. 1; pp. 52-58, ISSN 1229-9197.

Kan, C. W.; Chan, K.; Yuen, C. W. M. & Miao, M. H. (1998). Plasma Modification on Wool Fibre: Effect on the Dyeing Properties, *J. Soc. Dyers Colour.* Vol. 114, pp. 61–65, ISSN 0037-9859.

Kan, C. W. (2006). Dyeing Behavior of Low Temperature Plasma Treated Wool, *Fibers Polym.* Vol. 7, No. 3, pp. 262–269, ISSN 1229-9197.

Kan, C. W. & Yuen, C. W. M. (2008). *Plasma Technology in Wool, (Textile Progress).* CRC Press, Vol. 39, pp. 121–187 ISBN: 9780415467452, United States.

Kerscher, M.,Korting, H.C. & Scharfer-Korting, M. (1991). Skin Ceramides: Structure and Function. *Eur. J. Dermatol.*, Vol. 1, pp. 1-25, ISSN 1167-1122.

Klausen, T.; Thomas, H. & Höcker, H. (1995). Influence of Oxygen Plasma Treatment in the Chemical and Morphological Changes of the Wool Fiber Surface. *Proceedings 9th Int. Wool Text. Res. Conf.* Biella, International Wool Textile Organization, Vol. 2, pp. 241-248.

Körner, A. & Wortmann, G. (2005). Isolation of 18-MEA Containing Proteolipids from Wool Fiber Cuticle, *Proceedings 11th Int. Wool Text. Res. Conf.* Leeds, University of Leeds, Vol. 92FWSA, ISBN 0-9553154-0-9.

Körner, A.; Petrovic, S. & Höcker, H. (1995). Cell Membrane lipids of Wool and Human Hair Form Liposomes. *Textile Res. J.*, Vol. 65, No. 1, pp. 56-58, ISSN 1746-7748.

Lasic D.D. (1993). Liposomes: From Physics to Application, Elsevier, Amsterdam, ISSN 978-0444895486

Lee, M.; Wakida, T.; Lee, M. S.; Pak, P. K. & Chen, J. (2001). Dyeing Transition Temperature of Wools Treated with Low Temperature Plasma, Liquid Ammonia, and High-pressure Steam in Dyeing with Acid and Disperse Dyes. *J. Appl. Polym. Sci.* Vol. 80, pp. 1058–1062, ISSN 1097-4628.

Leeder J.D. (1999). Comments on "Pathways for Aqueous Diffusion in Keratin Fibers", *Textile Res. J.,* Vol. 69, No.3, pp. 229, ISSN 1746-7748.

Leeder, J.D.; Holt, L.A.; Rippon, J.A. & Stapleton, I. W. (1990). Diffusion of Dyes and Other Reagents into the Wool Fibre. *Proceedings 8th Int. Wool Text. Res. Conf.,* Christchurch, Vol. 4, pp. 227-238.

Leeder, J.D. & Rippon, J.A. (1982). Histological Differentiation of Wool Fibres in Formic Acid. *J. Textile Institute.,* Vol. 73, No.3, pp. 149-151, ISSN 0040-5000.

Leeder, J.D.; Rippon, J.A. & Rivett, D.E. (1985a). Modification of the Surface Properties of Wool by Treatment with Anhydrous Alkali. *Proceedings 7th Int. Wool Text. Res. Conf.,* Tokyo, Vol.4, pp.312-321.

Leeder, J.D.; Rippon, J.A.; Rothery, F.E. & Stapleton, I.W. (1985b). Use of the Transmission Electron Microscope to study Dyeing and Diffusion Processes. *Proceedings 7th Int. Wool Text. Res. Conf.,* Tokyo, Vol.5, pp.99-108.

Lindberg, J.; Mercer, E.H.; Philip, B. & Gralén, N. (1949). The Fine Hystology of the Keratin Fibers. *Textile Res J.,* Vol. 19, No. 11, pp. 673-678, ISSN 1746-7748.

Lindberg, J. (1953). Relationship between Various Surface Properties of Wool Fibers: Part III: Sorption of Dyes and Acids in Wool Fibers. *Textile Res J.,* Vol. 23, pp. 573-584, ISSN 1746-7748.

Makinson, K. R. (1968). Some New Observations on the Effects of Mild Shrinkproofing Treatments on Wool Fibers. *Textile Res J.,* Vol. 38, No. 8, pp. 831-842, ISSN 1746-7748.

Manich, A.M.; Carilla, J.; Vilchez, S.; de Castellar, M.D.; Oller, P. & Erra, P. (2005). Thermomechanical Analysis of Merino Wool Yarns. *J. Therm. Anal. Calorim.* Vol. 82, No.1, pp.119-123, ISSN: 1388-6150.

Martí, M.; Serra, S.; de la Maza, A.; Parra, J.L. & Coderch, L. (2001). Dyeing Wool at Low Temperatures: New Method Using Liposomes. *Textile Res. J.,* Vol. 71, No.8, pp. 678-682 ISSN 1746-7748.

Martí, M.; Coderch, L.; de la Maza, A.; Manich, A.M. & Parra J.L. (1998). Phosphatidilcholine Liposomes as Vehicles for Disperse Dyes for dyeing Polyester/Wool Blends. *Textile Res. J.,* Vol.68, No. 3, pp. 209218, ISSN 1746-7748.

Martí, M.; Barsukov, L.I.; Fonollosa, J.; Parra, J.L.; Sukhanov, S.V. & Coderch, L. (2004). Physicochemical Aspects of the Liposome-Wool Interaction in Wool Dyeing. *Langmuir.* Vol. 20, pp. 3068-3073, ISSN 0743-7463.

Martí, M.; Ramírez, R.; Manich, A.M.; Coderch, L. & Parra J.L. (2007). Thermal Analysis of Merino Fibers without Internal Lipids. *J. of App. Polym. Sci.,* Vol. 104, No. 1, pp. 545-551, ISSN 1097-4628.

Martí, M.; Ramirez, R.; Barba, C.; Coderch, L. & Parra, J.L. (2010). Influence of Internal Lipid on Dyeing of Wool Fibers. *Textile Res. J.* Vol. 80, No. 4, pp. 365-373, ISSN 1746-7748.

Martí, M.; Manich, A.M.; Oller, P.; Coderch, L.; Parra, J.L. (2005). Calorimetric Study of Lipid Extracted Wool Fibers. *Proceedings 11th Int. Wool Res. Conf.,* Leeds. Fi 5S-3, ISBN 0-9553154-0-9.

Martí, M.; de la Maza, A.; Parra, J.L. & Coderch, L. (2001). Dyeing wool at low temperatures: new method using liposomes. *Textile Res J.*, Vol. 71, No. 8, pp. 678–682, ISSN 1746-7748.

Medley, J.A. & Andrews, M.W. (1959). The Effect a Surface Barrier on Uptake Rates of Dyes into Wool Fiber. *Textile Res. J.* Vol.29, No. 5, pp. 398-403, ISSN 1746-7748.

Méndez, S.; Martí, M.; Barba, C.; Parra, J.L. & Coderch, L. (2007). Thermotropic behavior of ceramides and their isolation from wool. *Langmuir*, Vol. 23, No. 3, pp. 1359-1364, ISSN 0743-7463.

Millson, H.E. & Turl, L.H. (1950). *Amer. Dyestuff Reptr.* Vol.93, pp. 647.

Montazer, M.; Validi, M. & Toliyat, T. (2006). Influence of temperature on stability of multilamellar liposomes in wool dyeing. *J. Liposome Res.* Vol 16, No.1, pp. 81-89, ISSN 0898-2104.

Montazer, M.; Taghavi, F.A.; Toliyat, T. & Bameni Moghadam, M. (2007). Optimizing of Dyeing of Wool with Madder and Liposomes by Central Composite Design. *J. of Appl. Polym. Sci*, Vol. 106, No. 4, pp.1614-1621, ISSN 1097-4628.

Naebe, M.; Cooksoon, PG.; Rippon, J.; Brady, R.P. & Wang, X. (2010). Effects of Plasma Treatment of Wool on the Uptake of Sulfonated Dyes with Different Hydrophpbic Properties. *Textile Res. J.* Vol. 80, No. 4, pp. 312-324, ISSN 1746-7748

Negri, A.P.; Cornell, H.J. & Rivett, D.E. (1993). The modification of the Surface Diffusion Barrier Wool. *J. Soc. Dyers Colour.* Vol. 109, No. 9, pp.296-301, ISSN 0037-9859.

Petersen, R.D. (1992). Ceramides: Key Components for Skin Protection. *Cosmet. Toiletries.* Vol. 107, pp. 45-49, ISSN 0361-4387.

Ramírez, R.; Garay, I.; Álvarez, J.; Martí, M.; Parra, J.L. & Coderch, L. (2008a). Supercritical fluid extraction to obtain ceramides from wool fibers. *Sep. Purif. Technol.* Vol.63, pp. 552-557, ISSN 1383-5866.

Ramírez, R.; Martí, M.; Manich, A.M.; Parra, J.L. & Coderch, L. (2008b). Ceramides extracted from wool: pilot plant solvent extraction. *Textile Res. J.* Vol.78, No. 1, pp. 73-80, ISSN 1746-7748.

Ramírez, R.; Martí, M.; Garay, I.; Manich, A.M.; Parra, J.L. & Coderch, L. (2009b). Ceramides extracted from wool: Supercritical Extraction Processes. *Textile Res. J.* Vol.79, No. 8, pp. 721-727, ISSN 1746-7748.

Ramirez, R.; Martí, M.; Cavaco-Paulo, A.; Silva, R.; de la Maza, A.; Parra, J.L. & Coderch, L. (2009a). Liposome formation with wool lipid extracts rich in ceramides, *J. Liposome Res.*, Vol. 19, pp. 77-83, ISSN 0898-2104.

Rippon, J. A. (1992). The Structure of Wool, in "Wool Dyeing," Lewis, D. M., (Ed.), Society of Dyers and Colourists, Bradford, England, pp. 1–51, ISBN 0 901956 53 8.

Rippon J.A. (1999). Comments on "Pathways for Dye Diffusion in Wool Fibers", *Textile Res. J.* Vol. 69, No. 4, pp. 307-308, ISSN 1746-7748.

Rippon, J.A. & Leeder, J.D. (1986). The Effect of Treatment with Percholoethylene on the Abrasion Resistance of Wool Fabric. *J. Soc. Dyers Colour.* Vol. 102, pp. 171-176, ISSN 0037-9859.

Rivett, D.E. (1991). Structural Lipids of the Wool Fibre. Wool Sci. Rev., 67, 1-25

Rocha Gomes, J.I.N.; Genovez, M.C. & Hrdina, R. (1997). Controlling exhaustion of reactive dyes on wool by microencapsulation with liposomes. *Textile Res J.*, Vol. 67, pp. 537–541, ISSN 1746-7748.

Schaefer, H. & Redelmeier, T.E. (1996). Skin Barrier: Principles in Percutaneous Penetration pp. 55-58, Karger, ISBN 3-8055-6326, Basel, Switzerland.

Schürer, N.Y.; Plewing, G.; & Elias P.M. (1991). Stratum Corneum Lipid Function. *Dermatologica.* Vol.183, No. 2, pp. 77-94, ISSN 1018-8665.

Simmonds, D.H. (1955). *Aust. J. Biol. Sci.* Vol.8, pp. 537.

Simonova, T.N.; Sukhanov, S.V.; Barsukova, O.V. & Barsukov, L.I. (2000). Liposomes as a Tool in studies of Wool dyeing. *Proceesings of the 10th International Wool Textile Research Conference,* Deutsches Wollforschungsinstitut: Aachen, Germany, 2000, DY-8, pp.121, ISBN 3-00-007905-X.

Spei, M. & Holzem, R. (1987). Thermoanalytical Investigations of Extended and Annealed Keratins. *Colloid & Polym. Sci.* Vol. 265, pp.965-970, ISSN 0303-402X.

Swift J.A. (1999). Pathways for Aqueous Diffusion in Keratin Fibers. *Textile Res. J.* Vol. 69, No. 2, pp. 152, ISSN 1746-7748.

Thomas, H.; Lehmann, G.; Höcker, H.; Möller, M.; Thode, C. & Lindmayer, M. (2005). Plasma Pre-Treatment at Atmospheric Conditions for Improved Processing and Performance Characteristics of Wool Fabrics. *Proceedings 11th Int. Wool Text. Res. Conf.* Leeds, University of Leeds, 86 FWS, ISBN 0-9553154-0-9.

Thomas, H. (2007). Plasma Modification of Wool, In: *Plasma Technologies for Textiles*, Shishoo, R., (Ed.), pp. 228–246, Woodhead Publishing Limited, Cambridge, England.

Wakida, T.; Tokino, S.; Niu, S.; Kawamura, H.; Sato, Y.; Lee, M.; Uchiyama, H. & Inagaki, H. (1993). Surface Characteristics of Wool and Poly(ethylene Terephthalate) Fabrics and Film Treated with Low Temperature Plasma Under Atmospheric Pressure, *Textile Res. J.* Vol.63, No. 8, pp.433–438, ISSN 1746-7748.

Wakida, T.; Lee, M.; Sato, Y.; Ogasawara, S.; Ge, Y. & Niu, S. (1996). Dyeing Properties of Oxygen Low-temperature Plasmatreated Wool and Nylon 6 Fibres with Acid and Basic Dyes. *J. Soc. Dyers Colour.* Vol. 112, pp.233-236, ISSN 0037-9859.

Wakida, T.; Lee, M.; Niu, S.; Kobayashi, S. & Ogasawara, S. (1994). Microscopic Observation of Cross-section of Dyed Wool and Nylon 6 Fibres after Treatment with Low-temperature Plasma. *Seni Gakkaishi* Vol. 50, pp. 421–423.

Ward, R.J.; Willis, H.A.; George, G.A.; Guise, G.B.; Denning, R.J.; Evans, D.J. & Short, R.D. (1993). Surface Analysis of Wool by X-Ray Photoelectron Spectroscopy and Static Secondary Ion Mass Spectrometry. *Textile Res J.*, Vol. 63, No. 6, pp. 362-368, ISSN 1746-7748.

Wortmann, F.J.; Wortmann, G. & Zahn, H. (1997). Pathways for Dye Diffusion in Wool Fibers, *Textile Res. J.*, Vol. 67, pp. 720-724, ISSN 1746-7748.

Wortmann, F.J. & Deutz, H.J. (1998). Thermal Analysis of Ortho- and Para-cortical Cells Isolated from Wool Fibers. *J. Appl. Polym. Sci.* Vol. 68, No. 12, pp. 1991-1995, ISSN:1097-4628.

Wortmann, F.J. (2005). Facets of the Structure and Physical Properties of Keratin Fibres. *Proceedings of the 11th Intern. Wool Res. Conf.* Leeds, September 2005, 19 KN, ISBN 0-9553154-0-9.

Wortmann, F.J. & Deutz, H.J. (1993). Characterizing Keratin Using High-pressure Differential Scanning Calorimetry (HPDSC). *J. Appl. Polym. Sci.* Vol. 48, No. 1 pp. 137-150, ISSN 1097-4628.

Yoon, N. S.; Lim, Y. J.; Tahara, M. & Takagishi, T. (1996). Mechanical and Dyeing Properties of Wool and Cotton Fabrics Treated with Low Temperature Plasma and Enzymes, *Textile Res. J.* Vol. 66, No. 5, pp. 329–336, ISSN 1746-7748.

Zarubina, N.P.; Belokurova, O.A. & Telegin, F.Y. (2000). Thermal Transition in Wool Evaluated by Sorption of Acid Dyes. *Proceedings 10th Int. Wool Text. Res. Conf.* Aachen, German Wool Research Institute, Vol. ST-P9, pp. 1-8, ISBN 3-00-007905-X

Zhan, H. (1980). Plenary Lecture, 6th *Intern. Wool Res. Conf.* Pretoria, Vol. 1

5

Natural Dye from Eucalyptus Leaves and Application for Wool Fabric Dyeing by Using Padding Techniques

Rattanaphol Mongkholrattanasit[1], Jiří Kryštůfek[2],
Jakub Wiener[2] and Jarmila Studničková[2]
*[1]Department of Textile Chemistry Technology, Faculty of Industrial Textile and Fashion
Design, Rajamangala University of Technology Phra Nakhon, Bangkok,
[2]Department of Textile Chemistry, Faculty of Textile Engineering,
Technical University of Liberec, Liberec,
[1]Thailand
[2]Czech Republic*

1. Introduction

Natural dyes are known for their use in colouring of food substrate, leather, wood as well as natural fibers like wool, silk, cotton and flax as major areas of application since ancient times. Natural dyes have a wide range of shades that can be obtained from various parts of plants, including roots, bark, leaves, flowers and fruits (Allen, 1971). Since the advent of widely available and cheaper synthetic dyes in 1856 having moderate to excellent colour fastness properties, the use of natural dyes having poor to moderate wash and light fastness has declined to a great extent. However, recently there has been revival of the growing interest on the application of natural dyes on natural fibers due to worldwide environmental consciousness (Samanta & Agarwal, 2009). Although this ancient art of dyeing with natural dyeing with natural dyes withstood the ravages of time, a rapid decline in natural dyeing continued due to the wide available of synthetic dyes at an economical price. However, even after a century, the use of natural dyes never erodes completely and they are still being used. Thus, natural dyeing of different textiles and leathers has been continued mainly in the decentralized sector for specialty products along with the use of synthetic dyes in the large scale sector for general textiles owing to the specific advantages and limitations of both natural dyes and synthetic dyes. The use of non-toxic and ecofriendly natural dyes on textiles has become a matter of significant importance because of the increased environmental awareness in order to avoid some hazardous synthetic dyes. However, worldwide the use of natural dyes for the colouration of textiles has mainly been confined to craftsman, small scale dyers and printers as well as small scale exporters and producers dealing with high valued ecofriendly textile production and sales (Samanta & Agarwal, 2009; Bechtold & Mussak, 2009; Vankar, 2007). Recently, a number of commercial dyers and small textile export houses have started looking at the possibilities of using natural dyes for regular basis dyeing and printing of textiles to overcome environmental pollution caused by the synthetic dyes (Glover & Pierce, 1993). Natural dyes produce very uncommon, soothing and soft shades as compared to synthetic dyes. On the other hand, synthetic dyes are widely available at an economical price and produce a wide variety of colours; these dyes however produce skin allergy, toxic wastes and other harmfulness to human body. There are a small number of companies that are known to produce natural dyes commercially. For example, de la Robbia, which began in 1992 in Milan, produces water extracts of natural dyes such as weld, chlorophyll, logwood, and cochineal under the Eco-Tex certifying system, and supplies the textile industry. In USA, Allegro Natural Dyes produces natural dyes under the Ecolour label for textile industry (Hwang et al., 2008). Aware of the Toxic Substance Act and the Environmental Protection Agency, they claim to have developed a mordant using a non-toxic aluminium formulation and biodegradable auxiliary substance. In Germany, Livos Pflanzenchemie Forschungs and Entwicklungs GmbH marked numerous natural products. In France, Bleu de Pastel sold an extract of woad leaves. Rubia Pigmenta Naturalia is The Netherlands company, which manufactures and sells vegetable dyes. There are several small textile companies using natural dyes. India is still a major producer of most natural dyed textiles (Vankar, 2007). Production of synthetic dyes is dependent on petrochemical source, and some of synthetic dyes contain toxic or carcinogenic amines which are not ecofriendly (Hunger, 2003). Moreover, the global consumption of textiles is estimated at around 30 million tonnes, which is expected to grow at the rate of 3% per annum. The colouration of this huge quantity of textiles needs around 700,000 tonnes of dyes which causes release of a vast amount of unused and unfixed synthetic colourants into the environment (Samanta & Agarwal, 2009). This practice cannot be stopped, because consumers always demand coloured textiles for eye-appeal, decoration and even for aesthetic purposes. Moreover, such a huge amount of required textiles materials cannot be dyed with natural dyes alone. Hence, the use of eco-safe synthetic dyes is also essential. But a certain portion of coloured textiles can always be supplemented and managed by eco-safe natural dyes (Samanta & Agarwal, 2009; Vankar, 2007). However, all natural dyes are not ecofriendly. There may be presence of heavy metals or some other form of toxicity in natural dye. So, the natural dyes also need to be tested for toxicity before their use (Vankar, 2007).

2. Natural organic dyes from eucalyptus

Eucalyptus is a members of evergreen hardwood genus, endemic to Australian. There are approximately nine hundred species and sub-species. Eucalyptus has also been successfully grown in many parts of the world, including southern Europe, Asia and the west coast of the United States (Flint, 2007). Eucalyptus is one of the most important sources of natural dye that gives yellowish-brown colourants. The colouring substance of eucalyptus has ample natural tannins and polyphenols varying from 10% to 12% (Ali et al., 2007). The major colouring component of eucalyptus bark is quercetin, which is also an antioxidant. It has been used as a food dye with high antioxidant properties (Vankar et al., 2006). Eucalyptus leaves contain up to 11% of the major components of tannin (gallic acid [3,4,5 – trihydroxybenzoic acid], with ellagic acid [2,3,7,8-tetrahydroxy (1) benzopyrano (5,4,3-cde) (1) benzopyran-5,10-dione]) and flavonoids (quercetin [3,3′,4′,5,7-pentahydroxylflavone] and rutin 3,3′,4′,5,7-pentahydroxylflavone-3-rhamnoglucoside]) as the minor components (Chapuis-Lardy et al., 2002; Conde et al., 1997). The structures of the colouring components found in eucalyptus leaves are given in Fig. 1. Tannins and flavonoids are considered very useful substances during the dyeing process because of their ability to fix dyes within fabrics. Silk dyed with an aqueous extract of eucalyptus leaves and bark possessing a mordant compound displays a yellowish-brown colour. An exception was when the fabric was dyed with ferrous mordant, resulting in a shade of dark brownish-grey. Colour fastness to water, washing, and perspiration was at good to very good levels, whereas colour fastness to light and rubbing exhibited fair to good levels (Mongkholrattanasit et al., 2007; Mongkholrattanasit et al., 2010; Mongkholrattanasit et al., 2011).

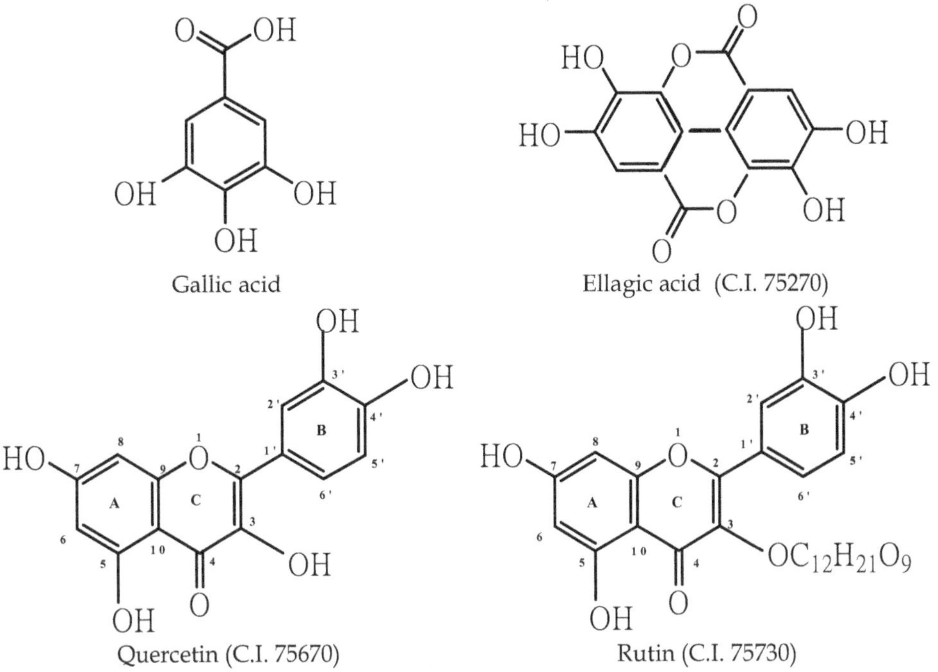

Gallic acid

Ellagic acid (C.I. 75270)

Quercetin (C.I. 75670)

Rutin (C.I. 75730)

Fig. 1. Colour composition of eucalyptus leaf extract dye

3. Using of natural dyes

Currently, application of natural dye incorporates new technology not only to exploit traditional techniques but also to improve the rate, cost and consistency production. It therefore, requires some special measurement to ensure evenness in dyeing. The processes of natural dyes for textile dyeing are as follows:

3.1 Extraction

Efficient extraction of the dyes from plant material is very important for standardization and optimization of vegetable dyes, utilizing a) soxhlet b) supercritical fluid extraction c) subcritical water extraction and d) sonicator method.

3.2 Dyeing

Normally, one technique used for dyeing with natural dye; exhaustion dyeing (conventional dyeing, sonicator dyeing and microwave dyeing). Exhaustion dyeing is using lot of water as shown in "Liquor Ratio (ratio between water and goods)". Producers immerge the goods in dye for extended periods for complete penetrate. This produces excessive waste water compared to a continuous process. The techniques used for dyeing of natural dyes, such as

1. Conventional dyeing : conventional dyeing is carried out by boiling the fabric in dye bath for 4-hours and often the dye uptake is still not completed. Enormous amount of heat is consumed in terms of heating the dye bath (Vankar, 2007).
2. Sonicator dyeing: utilization of ultrasound energy to aid wet processing of fabrics. The process of increasing dye transfer from the dye-bath to fabric using ultrasound energy is a function of the acoustic impedance characteristics of the fabrics (Vankar, 2007 ; Vankar et al., 2009 ; Tiwari & Vankar, 2001).
3. Microwave dyeing : microwave dyeing take into account only the dielectric and the thermal properties. The dielectric property refers to the intrinsic electrical properties which affect dyeing by dyeing by dipolar rotation of the dye and the influence of microwave field upon dipoles. The aqueous solution of dye has two components, which are polar. In the high frequency microwave field, oscillating at 2450 MHz; it influences the vibrational energy in the water molecule and the dye molecules (Tiwari & Vankar, 2001).

3.3 Mordanting

In the actual dyeing process, there are four ways of using mordant (Bechtold & Mussak, 2009; Moeyes, 1993) as follows:

a. Mordanting before dyeing, or pre-mordanting;
b. Mordanting and dyeing at the same time, called stuffing or simultaneous;
c. Mordanting after dyeing, or after-mordanting or post-mordanting;
d. A combination of pre-mordanting and after-mordanting.

4. Theoretical presuppositions of natural dyes to dyeing

Achieving a good, or at least a relatively good, water solubility using natural dyes is rather exceptional. No chemical group is capable of electrolytic dissociation or ionization in a molecule; an interesting and important exception is the anthocyanins, for example, pelargonidine, cyanidine, and betanidine are slightly cationic dyes and, therefore, also have relatively good solubility in water (Mongkholrattanasit et al., 2009). The "conditional solubility" of indigoid natural dyes, which in their original form are entirely insoluble, presents a quite special principle. In fact, indigo has been imitated to a great extent; synthetic indigo and their derivatives were produced on an industrial scale at the end of the nineteenth century as a forerunner of the latter large group of vat dyestuffs. The alkali reductive conversion of this fully insoluble compound in a proper soluble sodium salt of leucocompound with affinity to fibers and their oxidation after dyeing with the primary insoluble vat dye, which is finely dispersed in the fiber, is well known. What do the majority of natural dyes have in common? The chemical constitution (and corresponding physical properties) of indigo and other anthocyanin dyes has remarkable similarity with the modern synthetic disperse dyes: the solubility of more or less elongated molecules of chromogen is due to the presence of several polar groups (mainly –OH) on aromatic rings. No groups are capable of electrolytic ionization (with the exception of the anthocyanin and betanin). From this follows that they only have low solubility in water. Empirically, it is known that it is impossible to strengthen dyeing of cotton with natural dyes, but it can be done by adding neutral electrolytes (sodium chloride or sulfate) as substantive dyes. And bath acidifying, while having a significant effect on the so-called acid dyes (coloured sodium salts of sulfonic acids), has a negligible effect on the natural dyes.The structure of the flavonoid-colouring components of eucalyptus leaves and tannin (Fig. 1) is compared with the typical azo and anthraquinone disperse dye (Fig. 2).

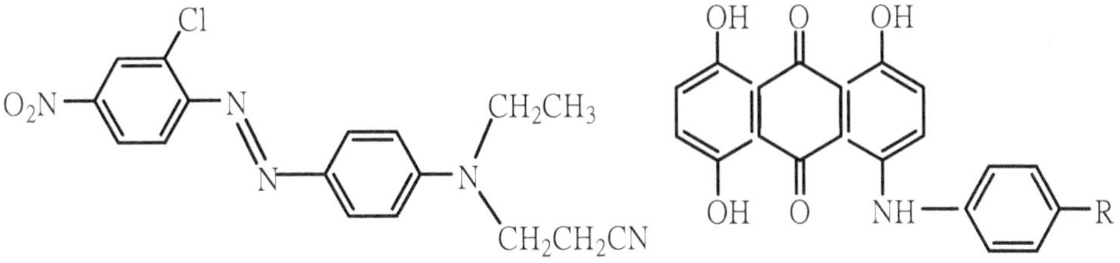

(a)

C.I. Disperse Red 50

(b)

R = H (C.I. Disperse Blue 27)
R = CH_3 (C.I. Disperse Blue 72)

Fig. 2. Chemical constitution of typical disperse dyes. (a) Azo dye and (b) anthraquinone dyes

Assume that most natural dyes are, on the basis of modern dyeing science, the disperse dyes. But what are the dyes for wool, silk, cotton, and flax? Consider that each fiber type in dyeing has already been studied, and it has become apparent that the disperse dyes are not good dyes for the aforementioned fibers. On the contrary, the synthetic disperse dyestuffs were developed for dyeing acetyl cellulose and synthetic fibers (i.e., hydrophobic fibers), and they have a low affinity for wool, silk, cotton, and other such fibers that are mainly hydrophilic. Though low, the indispensable affinity of disperse dyes makes them very undesirable for the staining of wool or cotton component by the dyeing of fiber mixtures, namely with polyester fiber (which is dyeable only in disperse dyes). This imperfect colouration-staining must be rather difficult to remove from wool or cotton component after dyeing because of its poor wet fastness and mostly unpleasant shade, which can be different from the shade of the same dye on polyester. However, the above-mentioned majority of natural dyes are providing only inexpressive wet fastness on wool and cotton fibers, and the mordanting by salts of suitable metals is also needed to improve wet fastness (not only to deepen but also to intensify the colour). A lower affinity results in the low dye exhaustion after the dye bath on the fiber. This can also be observed in the dyeing of natural fibers with natural dyes, such as the indigoid and anthocyanin dyes.

5. Ecological and economical aspects of dyeing with natural dyes

If we carry out the dyeing process with natural dyes in a slightly large manufacturing unit or a factory rather than in a household unit, we can surpass the limits of historic methods of dyeing and material pretreatments, which are lengthy and uneconomical procedures. The old methods (likely transmitted without facing critical evaluation), consist of various actions that do not address modern requirements, and do not take into account the new possibilities offered by the modern textile chemistry. The number and duration of baths seem to be too high (at least for European standards and customs) and are non-productive. For example, the required 3-5 hours wetting of material with water before dyeing could be greatly reduced by wetting in a bath by specially made wetting agent, and this or another agent could also be added into the dyeing bath. The ineffective use of natural dyes was already discussed above. The majority of dyes ceases as effluents in sewer. The mordanting salts do not have affinity to the fibers and therefore only a small part of them is bounded with fibers, and after dyeing and final rinsing all the remnants are carried off by water. What about the idea of storing the mordanting baths for future use? While logical, the number and volume of stock reservoirs (and place in dye house) make it an unpractical possibility. Naturally, serious conception-questions follow from this. Should "natural dyeing" remain as something principally untouchable whose traditional originality must be safe-guarded at any costs, or are we going to consider this natural raw-material source as an ecologically favorable supplement to synthetic colourants? or, can we synthesize the methodologies of "natural dyeing" with the research and application processes of modern dyeing technology? Nevertheless, both natural dyeing and modern dyeing technology can coexist. In any case, we are trying to explore the second of the following:

- the consequent minimization of concentration of natural dyes and mordants,
- the shortening of operating times, i.e., to save energy and productivity, and
- the maximal efficient use of dye and mordanting baths.

All these can be assured by the padding (pad) technologies, in which the liquor ratio (weight of textiles: bath) is about one order lower ($\leq 1:1$) than the common exhaustion (bath or batch) dyeing methods. The padding technologies are particularly advantageous to dyeing with the low-affinity products, because the dye affinity to fiber by padding is unnecessary (in phase of the dye deposition on the fabric). The dye bath is cloth "padded": mechanically applied by the rapid passage through the small padding trough, the intensive squeezing between expression rollers follows immediately. The process of padding is continuous and very rapid. It depends on the arrangement of the following dye fixation if the total procedure is continuous or semi continuous. The dye bath by padding is about one order higher than by the common dyeing from the "long bath" (the so-called exhaustion methods), in which the dyestuff exhausts on the fiber in consequence to its affinity to the fiber. The higher padding bath concentration results in more rapid dye diffusion in fiber during the next fixation operation. Much smaller bath volume (related to the fiber unit) causes the higher dye exploitation. In the case of natural dyes, the dye fixation is based on the reaction (see also Agarwal and Patel) (Agarwal & Patel, 2001) with the salts of complex-forming metals-mordants in the same or next bath-or the textile can be pre-metalized with mordant (this pre-mordanting is carried out from the long bath-the large non-effectiveness is mentioned above. Therefore, we also experimented with pad-dry and pad-batch principle at this operation). In semi continuous dyeing technology, several methods of dye fixation are known. The following two principles are important for our purpose:

a. fixation by drying, the so-called pad-dry method, the process is rapid but requires a reliably functional drying device (an excellently even -drying effect breadth-ways and cross-ways in the fabric is necessary, otherwise it may result in colour depreciation and unevenness),
b. fixation by batching of the padded goods at room or slightly increased temperature, now known as the pad-batch method. The padded and rolled goods are wrapped up in an airtight plastic sheet so that no selvedge drying occurs during storage, which lasts 8–24 hours.

After both dye fixation methods water rinsing follows repeatedly.

6. Experimental

The research focused on the properties of pad-dyeing techniques, we investigated the dyeing and ultraviolet (UV) protection properties of wool fabric using an aqueous extract of eucalyptus leaves as the natural dye. Different factors affecting dyeing ability were also thoroughly investigated.

The following laboratory-grade mordants were used: aluminium potassium sulfate dodecahydrate ($AlK(SO_4)_2.12H_2O$), ferrous (II) sulfate heptahydrate ($FeSO_4.7H_2O$), copper (II) sulfate pentahydrate ($CuSO_4.5H_2O$) and stannous chloride pentahydrate ($SnCl_2.5H_2O$). The anionic wetting agent, Altaran S8 (sodium alkylsulfate), and soaping agent, Syntapon ABA, were supplied by Chemotex Decin, Czech Republic.

The mordanting and dyeing processes were carried out in a two-bowl padding mangle machine (Mathis, Typ-Nr. HVF.69805). A drying machine (Mathis Labdryer, Typ-Nr. LTE-2992) was used for the drying of the dyed fabrics. A GBC UV/VIS 916 (Australia) spectrophotometer and a Datacolor 3890 were employed for the absorbance and colour strength measurements, respectively. The transmittance and ultraviolet protection factor (UPF) values were measured by a Shimadzu UV3101 PC UV-VIS-NIR scanning spectrophotometer in the 190 nm to2100 nm range.

Fresh eucalyptus leaves (*E. Camaldulensis)* were dried in sunlight for one month and crumbled using a blender, and then were used as the raw material for dye extraction, which was achieved by the reflux technique: 70 g of crumbled eucalyptus leaves was mixed with one liter of distilled water and refluxed for one hour. The dye solution was filtered, evaporated, and dried under reduced pressure using a rotary evaporator. The crude dye extract of the eucalyptus leaves was then crumbled with a blender and used for obtaining the standard calibration curve. The dilution of the eucalyptus leaf extract gives a relatively clear solution with a linear dependence on the concentration absorbance, an absorption peak (λ_{max}) at 262 nm (Yarosh et al., 2001). The concentration of 20 g/l was calculated from a standard curve between concentrations of eucalyptus leaf dye solutions versus absorbance at the wavelength mentioned.

The pre-mordanting methods, wool fabrics were immersed in each mordant solution with anionic wetting agent and padded on a two-bowl padding mangle at 80% pick up. Next, the mordanted sample was impregnated in each eucalyptus dye concentration. After padding for 2 seconds the samples were dried at 90°C for 5 minutes for a pad-dry technique. Under the cold pad-batch dyeing technique, the padded fabric was rolled on a glass rod with a plastic sheet wrapped around the rolled fabric. Then it was kept at room temperature for 24 hours. After the dyeing step, the samples were washed in 1 g/l of a soaping agent, Syntapon ABA, at 80°C for 5 minutes, then air dried at room temperature. For the simultaneous mordanting (meta-mordanting) method (i.e. dyeing in the presence of mordants), the fabrics were immersed in a bath containing a mordant and the dye extract at room temperature and padded on a two-bowl padding mangle at 80% pick up. The processing of pad-dry, pad-batch and soaping were the same as above mention. In the post-mordanting method, the fabrics were immersed in each eucalyptus dye concentration and without mordant, followed by padded on a two-bowl padding mangle at 80% pick up. Then the padded samples were padded by mordanting. Further processing was the same as described in the pre-mordanting method.

The colour strength (*K/S*) and CIELAB of the dyed samples were evaluated using a spectrophotometer (Datacolor 3890). All measured sample showed the maximum absorption wavelength (λ_{max}) value at 400 nm. The *K/S* is a function of colour depth and is calculated by the Kubelka-Munk equation, $K/S = (1-R)^2/2R$, where R is reflectance, K is the sorption coefficient, and S is the scattering coefficient.

6.1 Identification of crude eucalyptus extracted dye

The crude eucalyptus leaf extract dye was characterized by ultraviolet-visible spectroscopy. The crude extraction solution (50 mg/l) was prepared by dissolving in distilled water. The spectrophotometer was scanned from 190 nm to 820 nm to obtain the UV/Visible spectra.

The UV-vis spectrum of the crude eucalyptus leaf extract dye in an aqueous solution is presented in Fig. 3. The characteristic spectrum shows absorptions in the 205-210 nm and 250-270 nm regions. Absorption in the 205-210 nm region may be attributed to various chromophores, including the C=C bond of various compounds, the C=O bond of carbonyl compounds, and the benzene ring (probably from aromatic compounds) (Pretsch et al., 2000). Absorption in the 250-270 nm regions may be attributed to the electronic transitions of benzene and its derivatives, which may include various aromatic compounds such as phenolics (Pretsch et al., 2000). It can be observed from Fig. 3 that the dye can absorb radiations in the UV-C region (200-290 nm), the UV-B region (290-320) and the UV-A region (320-400) (Feng et al., 2007).

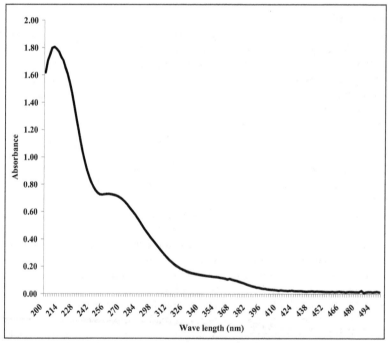

Fig. 3. UV-VIS spectrum of 50 mg/l crude eucalyptus leaf extract dye in distilled water.

6.2 Dyeing property of wool fabric dyed with eucalyptus leaf extract by using padding techniques by varying quantity of dye concentrations

The effect of mordanting methods and padding techniques on dyeing of wool fabric with different mordants are shown in Table 1 to Table 3. All measured sample showed the greatest

λ_{max} value at 400 nm. Table 1 to Table 3 show CIELAB L^*, a^*, b^* values for the wool fabric dyed with different mordants by three mordanting methods (pre-mordanting, simultaneous mordanting and post-mordanting) and using two padding techniques, namely the pad-batch and pad-dry techniques. L^*, a^*, b^* refer to the three axes of the CIELAB system. The L^* value indicates perceived lightness in CIELAB colour space. The L^* scale run from 0 (black) to 100 (white); the higher the L^* reading the lighter colour. The a^* value indicates red (+a^*) and green (-a^*) while the b^* value indicates yellow (+b^*) and blue (-b^*) (Sarkar & Seal, 2003 ; Giles,1974; Duff & Sinclair,1989). It can be observed that the K/S values increase with an increase of dye concentration. Little difference between the two padding techniques utilized for the wool fabric dyes by three mordanting methods, except wool fabrics mordanted with copper sulfate whose gave a high K/S values on the pad-batch technique than pad-dry technique. In all cases ferrous sulfate mordant yielded the best dyeing results, and the next good result was obtained in the order of copper sulfate, stannous chloride and alum. As observed from the K/S values, in the case of wool fabrics dyed with alum by using post-mordanting method gave lower colour strength than without mordant.

Type of mordant	Dye Conc. (g/l)	Pad-batch on wool fabric					Pad-dry on wool fabric				
		K/S	L^*	a^*	b^*	[1]Dyed sample	K/S	L^*	a^*	b^*	[1]Dyed sample
Without mordant	5	1.50	76.1	3.5	14.2		1.13	77.7	3.5	12.0	
	10	1.60	76.0	3.5	15.0		1.54	76.3	3.6	14.1	
	20	1.86	75.4	3.5	15.8		1.86	75.9	3.4	15.8	
AlK(SO$_4$)$_2$ (Al)	5	1.44	79.0	0.4	19.1		1.36	79.6	0.1	20.3	
	10	1.70	77.6	0.1	19.3		1.45	78.8	0.2	19.7	
	20	1.75	75.2	0.8	20.4		1.70	76.5	0.8	19.0	
CuSO$_4$ (Cu)	5	2.58	67.2	2.2	20.6		2.02	70.8	2.8	20.0	
	10	3.36	63.7	2.9	21.3		2.20	69.3	3.1	19.4	
	20	3.42	63.0	3.3	21.8		2.70	67.1	3.4	20.1	
FeSO$_4$ (Fe)	5	2.16	50.9	1.5	0.1		2.66	46.8	2.3	0.4	
	10	2.94	45.2	1.8	-0.6		3.52	43.1	3.1	0.6	
	20	4.22	40.0	2.2	-0.6		4.34	38.6	3.4	0.7	
SnCl$_2$ (Sn)	5	1.86	84.5	-0.4	26.2		2.14	83.0	-0.3	26.9	
	10	2.28	83.7	-0.3	28.4		2.25	83.0	-0.4	26.9	
	20	2.98	81.5	0.5	30.8		2.37	81.6	0.1	26.7	

Note: [1] 20g/l dye concentration

Table 1. Colour value of wool fabric dyed with eucalyptus leaf extract by pre-mordanting and padding techniques, with using 10 g/l of metal mordants at different concentration of the dye

Alum and ferrous sulfate were the best mordant during simultaneous mordanting method of dyeing. However, copper sulfate showed the best mordant during simultaneous mordanting and pre-mordanting method of dyeing. For the K/S value on dyed wool fabrics were only little different using stannous chloride as mordant during three mordanting methods. Wool dyed without mordant showed yellowish-brown shade. The samples mordanted with copper sulfate, stannous chloride, and alum produced medium to dark grayish-brown, bright yellow and pale yellow shades, respectively. With ferrous sulfate, the colour was darker and duller. This may be associated with a change of ferrous sulfate into a ferric form by reacting with oxygen in the air. Ferrous and ferric forms coexisted on the fibers and their spectra overlapped, resulting in a shift of λ_{max} and consequent colour change to a darker shad (Hwang, 2008). Additional, the tannins combined with ferrous salts to form complexes, which also result in a darker shade of fabric (Vankar, 2007). From the results, it can be postulated that wool fabric can be successfully dyed with eucalyptus leaf extract. This may be attributed to the fact that eucalyptus leaves are rich tannin (Conde et al., 1997), which are phenolic compounds that can form hydrogen bonds with carboxyl groups in the protein fibers (Agarwal & Patel, 2001).

Type of mordant	Dye Conc. (g/l)	Pad-batch on wool fabric					Pad-dry on wool fabric				
		K/S	L^*	a^*	b^*	[1]Dyed sample	K/S	L^*	a^*	b^*	[1]Dyed sample
Without mordant	5	1.50	76.1	3.5	14.2		1.13	77.7	3.5	12.0	
	10	1.60	76.0	3.5	15.0		1.54	76.3	3.6	14.1	
	20	1.86	75.4	3.5	15.8		1.86	75.9	3.4	15.8	
$AlK(SO_4)_2$ (Al)	5	1.65	78.2	0.4	23.6		1.65	79.4	0.1	27.2	
	10	1.91	76.8	0.1	24.0		1.81	78.0	0.1	27.6	
	20	2.55	74.9	0.7	22.9		2.60	74.5	1.0	27.2	
$CuSO_4$ (Cu)	5	3.27	63.8	0.04	21.1		2.32	65.5	2.0	19.6	
	10	3.44	62.6	1.0	21.2		2.62	64.0	2.8	19.5	
	20	4.12	59.6	2.2	21.1		2.80	62.5	3.3	19.0	
$FeSO_4$ (Fe)	5	3.93	40.1	1.3	-1.3		4.54	40.6	1.3	-1.0	
	10	4.23	40.0	1.1	-0.9		4.81	37.1	1.3	-1.0	
	20	4.62	38.5	1.2	-1.1		5.14	37.2	1.0	-0.9	
$SnCl_2$ (Sn)	5	2.15	83.8	-0.2	28.5		1.58	84.7	-0.9	22.5	
	10	2.38	84.1	-0.7	28.3		2.01	83.5	-0.4	27.4	
	20	2.67	83.4	-0.8	30.4		2.92	81.5	0.4	30.4	

Note: [1] 20g/l dye concentration

Table 2. Colour value of wool fabric dyed with eucalyptus leaf extract by simultaneous mordanting and padding techniques, with using 10 g/l of metal mordants at different concentration of the dye

Type of mordant	Dye Conc. (g/l)	Pad-batch on wool fabric					Pad-dry on wool fabric				
		K/S	L^*	a^*	b^*	[1]Dyed sample	K/S	L^*	a^*	b^*	[1]Dyed sample
Without mordant	5	1.50	76.1	3.5	13.7		1.13	77.7	3.5	12.0	
	10	1.60	76.0	3.5	15.1		1.54	76.3	3.6	14.1	
	20	1.86	75.4	3.5	15.7		1.86	75.9	3.4	15.8	
$AlK(SO_4)_2$ (Al)	5	1.11	81.3	-1.2	19.2		0.93	82.1	0.6	16.9	
	10	1.23	80.2	-1.0	20.1		1.10	80.8	0.4	18.5	
	20	1.39	79.3	-0.8	21.4		1.28	79.7	0.2	19.9	
$CuSO_4$ (Cu)	5	2.50	66.8	-1.1	19.5		1.84	68.5	0.4	17.3	
	10	2.81	65.4	-0.1	20.3		2.12	66.8	0.6	18.7	
	20	3.06	63.7	0.2	20.2		2.87	63.1	1.7	20.2	
$FeSO_4$ (Fe)	5	3.18	53.0	1.8	8.4		2.28	55.8	1.5	6.1	
	10	3.35	50.2	2.0	8.5		2.71	51.1	1.5	3.9	
	20	3.86	46.7	1.5	2.7		3.13	45.7	1.6	1.0	
$SnCl_2$ (Sn)	5	1.54	85.4	-0.3	25.6		1.52	86.6	-0.3	24.8	
	10	1.90	85.0	-0.2	27.8		1.88	85.1	-0.2	25.7	
	20	2.05	84.0	-0.1	28.1		2.01	84.1	-0.2	26.6	

Note: [1] 20g/l dye concentration

Table 3. Colour value of wool fabric dyed with eucalyptus leaf extract by post-mordanting and padding techniques, with using 10 g/l of metal mordants at different concentration of the dye

6.3 Effect of quantity of mordant concentrations, time/ temperature on pad-dry and batching time on pad-batch

Table 4 shows the colour values of wool fabric dyed with eucalyptus leaf extract by varying quantity of mordant concentrations. All measured sample showed the greatest λ_{max} value at 400 nm. It can be seen that the K/S values increase with an increase of mordant concentration. The dyed uptake values were greater at the higher mordant concentration. This could be attributed to the darkening and dulling of shades due to mordant effect. Little different between the two padding techniques utilized for the study is observed. Wool fabric dyed with eucalyptus leaf extract in the absence mordant showed yellowish brown shades. Comparison of four metal mordants showed that the ferrous sulfate metal mordant gave the highest depth of shade on wool fabric. Thus ferrous sulfate was the best mordant during mordanting method of dyeing. This could be attributed to difference in CIELAB values of the dyed samples. The mordant activity of the five sequences was as follows: Fe > Cu > Al > Sn > without mordanted in wool fabric, the absorption of colour by wool fabric was enhanced by using metal mordants.

Type of mordant	Conc. (g/l)	Pad-batch on wool fabric					Pad-dry on wool fabric				
		K/S	L^*	a^*	b^*	[1]Dyed sample	K/S	L^*	a^*	b^*	[1]Dyed sample
Without mordant	-	1.86	75.4	3.5	15.8		1.86	75.9	3.4	15.8	
$AlK(SO_4)_2$ (Al)	5	2.18	76.2	0.5	28.2		2.09	77.8	0.8	28.0	
	10	2.55	74.9	0.7	29.9		2.60	74.5	1.0	29.1	
	20	3.91	72.8	1.0	32.1		3.97	72.0	1.2	31.6	
$CuSO_4$ (Cu)	5	3.84	61.3	2.4	20.9		3.80	62.4	2.3	20.2	
	10	4.12	59.6	2.7	21.1		4.08	60.0	3.3	19.0	
	20	5.00	54.4	2.8	23.1		4.87	55.7	3.2	20.1	
$FeSO_4$ (Fe)	5	4.81	40.0	1.0	-0.9		4.95	39.1	1.1	-0.6	
	10	4.98	38.5	1.2	-1.1		5.14	37.0	1.0	-0.9	
	20	7.28	36.9	0.7	-1.4		7.60	36.0	0.9	-0.7	
$SnCl_2$ (Sn)	5	2.64	83.3	-1.6	32.2		2.66	83.1	-2.3	31.8	
	10	2.67	83.4	-1.2	30.4		2.71	82.5	-2.4	30.4	
	20	3.11	81.8	-2.1	35.3		3.13	81.4	-2.7	34.6	

Note: [1] 20g/l metal mordants concentration

Table 4. Colour value of wool fabric dyed with eucalyptus leaf extract by simultaneous mordanting and padding techniques, with using 20 g/l of dye concentration at different concentration of the mordant

From the results, it is clear that ferrous sulfate and copper sulfate mordants are well known for their ability to form coordinate complexes and in this experiment both readily chelated with the dye. As the coordination numbers of ferrous sulfate and copper sulfate are 6 and 4 respectively, some co-ordination sites remained unoccupied when they interacted with the fiber. Functional groups such as amino and carboxylic acid groups on the fiber can occupy these sites. Thus this metal can form a ternary complex on one site with the fiber and on the other site with the dye (Bhattacharya & Shah, 2000). Stannous chloride and alum metals formed weak coordination complexes with the dye, they tend to form quite strong bonds with the dye but not with the fiber, so they block the dye and reduce the dye interaction with the fiber (Bhattacharya & Shah, 2000).

The effect of time and temperature on colour strength (K/S) value was evaluated by padding a sample of wool fabric with eucalyptus leaf extract and ferrous sulfate as mordant. The samples were processed only by drying condition were 40°C, 60°C and 90°C for 1, 3, 5 and 10 minutes. The K/S values obtained are shown in Fig. 4. It is clear that the colour strength (K/S) values increase with in crease in the drying time and temperature in wool fabric. A study of Fig. 4 reveals that the high colour strength values (ca. 7.60) was achieved for the wool fabric on drying at 90°C for 5 minutes. The pad-batch dyeing process was carried out at room temperature with batching times of different lengths to assure an operation as economic as possible. Fig. 5 shows that low colour strength required a period of 1 hour, medium colour strength of 6-12 hours and high colour strength a period of 24 hours. The colour strength obtained was increased as the batching time increased for wool fabrics.

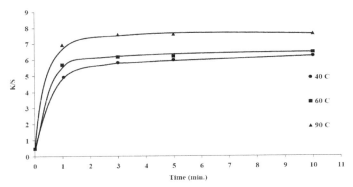

Fig. 4. Effect of drying time and temperature of pad-dry technique on the colour strength (K/S values) of wool fabric dyed with 20 g/l eucalyptus leaf extract and using 20 g/l ferrous sulfate by using simultaneous mordanting

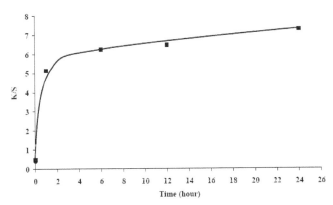

Fig. 5. Effect of batching time of pad-batch technique on the colour strength (K/S values) of wool fabric dyed with 20 g/l eucalyptus leaf extract and using 20 g/l ferrous sulfate by using simultaneous mordanting

6.4 The percentage yield (exploitation) of wool fabric dyed with eucalyptus leaf extract by simultaneous pad-dyeing

It was estimated that the best shades (deep and colour fastness) are obtained when mordanting with ferrous sulfate ($FeSO_4 \cdot 7H_2O$) and, therefore, this mordant was used for the experiments. The following concentration range of eucalyptus leaf extract and mordant $FeSO_4 \cdot 7H_2O$ in the same concentration was used: 1, 5, 10, 20, 30, and 40 g/l, and in all cases anionic wetting agent in the concentration of 1 g/l was added to the padding bath. Glacial acetic acid was added to maintain the pH of the liquid at 4. The simultaneous padding was carried out at room temperature in a two-bowl padding mangle using 80% pick up. After padding (2 seconds), the samples were dried at 90°C for 5 minutes and after 1 hour, all samples were repeatedly rinsed in warm water at 60°C until the rinsing water remained colourless. The rinsed water was collected with the rest of dyeing bath in the volumetric flask and filled up to the defined volume for absorbance measurement by UV–vis spectrophotometer (at the wavelength of 270 nm at which the maximum absorbance was recorded). The concentration of eucalyptus leaf extract fixed in the fiber and percentage of its use (percentage of yield) from bath on fiber were calculated from the absorbance of the rinsing water by using the standard graph. Relationship between bath concentration and padding condition were calculated from Eq. (1) to Eq. (6) (Mongkholrattanasit et al., 2009). We assume when the initial dye concentration in the pad bath is C_0 (g/l). The quantity of dye transported by fabric is C_{pi} (mg/g)

$$C_{pi} = \frac{\% \, pick \, up}{100} \cdot C_0 \tag{1}$$

The concentration of dye in conjoined-water after rinsing can be expressed as:

$$C_r = \frac{Absorbance}{\varepsilon \cdot l} \tag{2}$$

where C_r = the concentration of dye in conjoined-water (mg/l), ε = absorption coefficient (l/mole.cm) and l = layer of solution (cm). Then the concentration of dye, which was stripped from material, is C_w (mg/g)

$$C_w = \frac{C_r}{1,000} \frac{V}{g} \tag{3}$$

where V = total volume after rinsing (ml) and g = weight of material (g). The concentration of dye absorbed on material, C_s (mg/g) was calculated as:

$$C_s = C_{pi} - C_w \tag{4}$$

The percentage of dye which stripped from the material can be shown as Eq. (5)

$$W = \frac{C_w \cdot 100}{C_{pi}} \tag{5}$$

where W = the percentage of dye which stripped from the material (%). And the percentage of exploitation of dye (yield), E (%) can be calculated as:

$$E = 100 - W \tag{6}$$

Wool fabric dyed with the water extract of eucalyptus leaves in the presence of the $FeSO_4$ mordant in the same padding bath shows a colour range of a brown grey shade to a dark grey shade. In Table 5, the results are presented. The yield (exploitation) of the colouring component of eucalyptus leaf extract in wool fabric is about 68%–52% from the lowest to the highest concentrations, and this corresponds to the medium deep brown-grey shades in the concentrations of more than 20 g/l eucalyptus leaf extract.

C_0 (g/l)	Percentage of pick up	C_{pi} (mg/g)	C_s (mg/g)	Yield (%)	K/S value (400 nm)
1	80	0.8	0.5	68.0	1.8
5	80	4	2.5	62.8	2.8
10	80	8	4.2	53.2	3.7
20	80	16	8.3	52.0	3.9
30	80	24	12.6	52.6	4.0
40	80	32	16.0	52.2	4.5

Table 5. Percentage yield and K/S values obtained by the simultaneous pad-dyeing/ mordant of wool fabric

6.5 UV protection properties of wool fabric dyed with eucalyptus leaf extract

The transmittance and UPF values of the original wool fabric, and fabrics dyed with the eucalyptus leaf extract were measured using Shimadzu UV3101 PC (UV-VIS-NIR Scanning Spectrophotometer) in the range of 190 nm to 2100 nm. The UPF value of the fabric was determined from the total spectral transmittance based on AS/NZ 4399:1996 as follows (Gies et al., 2000)

$$UPF = \frac{\sum_{290}^{400} E_\lambda S_\lambda \Delta_\lambda}{\sum_{290}^{400} E_\lambda S_\lambda T_\lambda \Delta_\lambda}$$

where E_λ is the relative erythemal spectral effectiveness (unitless), S_λ is the solar ultraviolet radiation (UVR) spectral irradiance in $W.m^{-2}.nm^{-1}$, T_λ is the measured spectral transmission of the fabric, Δ_λ is the bandwidth in millimeter, and λ is the wavelength in nanometre. The UVR band consists of three regions: UV-A band (320 nm to - 400 nm), UV-B band (290 nm to 320 nm), and UV-C band (200 nm to 290 nm) (Feng et al., 2007). The highest energy region, the UV-C band, is absorbed completely by oxygen and ozone in the upper atmosphere. Of the solar UV radiation reaching the earth's surface, 6% is in the UV-B region and 94% in the UV-A region (Allen & Bain, 1994). UV-A causes little visible reaction on the skin but has been shown to decrease the immunological response of skin cells (Sarkar, 2003). UV-B is the most responsible for the development of skin cancers (Sarkar, 2003). Therefore, the transmittance of UVR, including UV-A and UV-B, through the fabrics was evaluated in this experiment. Fabrics with a UPF value in the range of 15-24 are defined as providing "good UV protection", 25-39 as "very Good UV protection", and 40 or greater as "excellent UV protection" (Sarkar, 2003). There is no rating assigned if the UPF value is greater than 50.

A commercially produced plain-weave wool fabric (thickness 0.36 mm, weight 193 g/m^2, fabric count per inch 62 x 54) was used in this experimental. The thread count, fabric thickness, and fabric weight characteristics of the wool fabric was in accordance with ASTM D3775-98, ISO 5084-1996, and ISO 3801-1997, respectively. A pre-mordanting padding process was used in this study.

To investigate the UV-protection property of eucalyptus leaves dye, UV transmittance spectra of the wool fabric with or without dyeing and the dyed wool fabric with mordants were compared. The percent UV transmittance data of wool fabric dyed with and without a mordanting agent are shown in Fig. 6. The results show significantly different between the dyed and undyed fabrics, which yields a high UV transmittance. The UV transmittance of the undyed wool was in the range of about 4-12% in the UV-B band and about 12-37% in the UV-A band. This indicates that the resistance of undyed fabrics to ultraviolet ray was very poor. While the UV transmittance of wool fabrics dyed by eucalyptus leaf extract appeared to be lower than 5% in the UV-B region. Generally, the UV protection property of fabrics is evaluated as good when the UV transmittance is less than 5% (Feng et al., 2007 ; Teng & Yu, 2003).

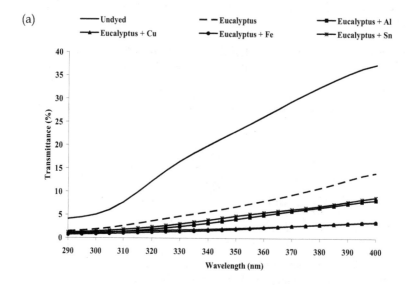

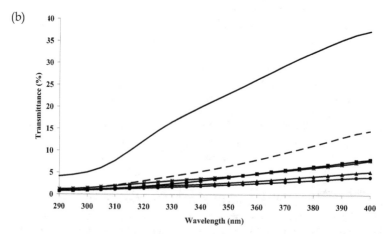

Note: Al = AlK(SO$_4$)$_2$, Cu = CuSO$_4$, Fe = FeSO$_4$, Sn = SnCl$_2$

Fig. 6. UV transmission of wool fabric dyed with 5 g/l eucalyptus leaf extract dye solution, , using 10 g/l mordants by (a) pad-batch and (b) pad-dry techniques.

For the samples mordanted with $AlK(SO_4)_2$, $CuSO_4$, $FeSO_4$ and $SnCl_2$, the percent UV-B transmittance was in the range of 0.8-1.9 %, 1.0-1.6%, 0.7-1.1%, and 1.2-2.7%, respectively for pad-batch and 0.7-1.8 %, 0.9-1.7%, 0.7-1.4%, and 1.2-2.5%, respectively for pad-dry. It is clearly seen that the values of the spectral transmittance are decreased with the mordants such as $AlK(SO_4)_2$, $CuSO_4$, $FeSO_4$, and $SnCl_2$ and different mordants had different effects on the spectral transmittance of the fabric dyed (Feng et al., 2007). Additionally, the colour and colour depth of the fabric can be related to UV transmittance in which light colours transmit more UV radiation than dark colours (Wilson et al., 2008). Table 6 shows the UPF values and protection class of wool fabric dyed by eucalyptus leaves with and without metal mordants by pad-dry and pad-batch dyeing techniques. Little difference is observed between the two padding techniques utilized for this study. The undyed fabric had a high transmittance and a very low UPF value of 10.8. The dyed samples without metal mordant in both dyeing techniques show UPF values between 32.8 and 35.4, which can be rated as offering Very good UV protection (UPF values between 25 and 39).

From the transmission data and the corresponding UPF values, all metal mordants used in this study caused a reduction in UV radiation transmission through the wool fabric. Wool fabric dyed by the metal mordants at 5 g/l concentrations of dye in the pad-dry and the pad-batch dyeing techniques could be classified as offering Excellent UV protection (UPF values 40 or greater). Wool fabrics, which after dyeing with and without mordant are rated as very good to excellent UV protection because wool fabric have low porosity and high weight and thickness. Therefore, wool fabric gives high UPF and permitting transmission of less UV radiation.

Mordant	Dye Conc. (g/l)	Pad-batch		Pad-dry	
		UPF	UPF Protection class	UPF	UPF Protection class
-	Un-dyed	10.8	No Class	10.8	No Class
Without	5	32.8	Very good	35.4	Very good
$AlK(SO_4)_2$	5	59.0	Excellent	55.1	Excellent
$CuSO_4$	5	67.9	Excellent	65.0	Excellent
$FeSO_4$	5	85.3	Excellent	81.8	Excellent
$SnCl_2$	5	46.9	Excellent	45.5	Excellent

Table 6. UPF values, protection class, and K/S values of wool fabric dyed with 5 g/l eucalyptus leaf extract dye solution and using 10 g/l mordants

6.6 Effect of dyeing technique on fastness properties

The colour fastness to washing, light, perspiration, water and rubbing of the dyed samples was determined according to ISO 105-C06 A1S:1994, ISO 105-B02:1994, ISO 105 E04: 1994, ISO 105-E01: 1994 and ISO 105-X12:2001, respectively.

The fastness rating of wool fabric dyed with or without mordants at 20 g/l dye concentration is presented in Tables 7 through 11. When comparing the fastness rating of the samples dyed using the two padding techniques, it can be postulated that the pad-batch technique gives nearly the same fastness properties as the pad-dry technique. Table 7 indicates that the washing fastness ratings of wool fabric dyed with eucalyptus leaves were very good (4-5). However, the light fastness was only fair (3-4), as shown in Table 8. The colour fastness to rubbing is shown to be in range of 4 to 4-5 (good to very good), except for fabrics mordanted with $FeSO_4$, whose rating was only 3-4 (fair to good) when subjected to wet rubbing, as shown in Table 9. The rating obtained for colour fastness to water in term of the degree of colour change and colour staining were very good (4 to 4-5), as shown in Table 10. The colour fastness to perspiration in acid and alkaline solution of fabrics dyed with and without metal mordants are shown in range of 4 to 4-5 as seen in Tables 11. The good fastness properties of wool fabric dyed with eucalyptus leaf extract may be attributed to the fact that these dyes contain tannin, which may help covalent bond formation with the fiber, thereby resulting in good fixation on the fibrous material. Again, these tannins, having a phenolic structure, can form metal chelation with different mordants. Hence, after mordanting, these tannins are insoluble in water, ultimately improving washing, water, and perspiration fastness (Agarwal & Patel, 2001).

Fastness	Pad-batch					Pad-dry				
	Without	Al	Cu	Fe	Sn	Without	Al	Cu	Fe	Sn
Colour change	4	4-5	4-5	4-5	4-5	4-5	4-5	4-5	4-5	4-5
Colour staining										
-Acetate	4-5	4-5	4-5	4-5	4-5	4-5	4-5	4-5	4-5	4-5
-Cotton	4-5	4-5	4-5	4-5	4-5	4-5	4-5	4-5	4-5	4-5
-Nylon	4-5	4-5	4	4	4-5	4-5	4-5	4	4	4-5
-Polyester	4-5	4-5	4-5	4-5	4-5	4-5	4-5	4-5	4-5	4-5
-Acrylic	4-5	4-5	4-5	4-5	4-5	4-5	4-5	4-5	4-5	4-5
-Wool	4-5	4-5	4	4	4-5	4-5	4-5	4-5	4	4-5

Note: Al = $AlK(SO_4)_2$, Cu = $CuSO_4$, Fe = $FeSO_4$, Sn = $SnCl_2$

Table 7. Colour fastness to washing at 40°C (ISO 105-C06 A1S: 1994)

Pad-batch (Colour change)					Pad-dry (Colour change)				
Without	Al	Cu	Fe	Sn	Without	Al	Cu	Fe	Sn
3	3	3-4	4	3	3	3	3-4	3-4	3

Note: Al = $AlK(SO_4)_2$, Cu = $CuSO_4$, Fe = $FeSO_4$, Sn = $SnCl_2$

Table 8. Colour fastness to light (ISO 105-B02: 1994).

mordant	Colour staining							
	Pad-batch				Pad-dry			
	Warp direction		Weft direction		Warp direction		Weft direction	
	Dry	Wet	Dry	Wet	Dry	Wet	Dry	Wet
without	4-5	4-5	4-5	4-5	4-5	4-5	4-5	4-5
$AlK(SO_4)_2$	4-5	4-5	4	4-5	4-5	4-5	4-5	4-5
$CuSO_4$	4-5	4-5	4	4-5	4-5	4	4-5	4
$FeSO_4$	4	4	4	3-4	4	3-4	4	3-4
$SnCl_2$	4-5	4-5	4-5	4-5	4-5	4-5	4-5	4-5

Table 9. Colour fastness to rubbing (ISO 105-X12: 2001).

Fastness	Pad-batch					Pad-dry				
	Without	Al	Cu	Fe	Sn	Without	Al	Cu	Fe	Sn
Colour change	4-5	4-5	4-5	4-5	4-5	4-5	4-5	4-5	4-5	4-5
Colour staining										
-Acetate	4-5	4-5	4-5	4-5	4-5	4-5	4-5	4-5	4-5	4-5
-Cotton	4	4	4	4	4	4	4	4	4	4
-Nylon	4-5	4-5	4	4	4-5	4-5	4-5	4	4	4-5
-Polyester	4-5	4-5	4-5	4	4-5	4-5	4-5	4-5	4	4-5
-Acrylic	4-5	4-5	4-5	4-5	4-5	4-5	4-5	4-5	4-5	4-5
-Wool	4-5	4-5	4-5	4	4-5	4-5	4-5	4-5	4-5	4

Note: Al = $AlK(SO_4)_2$, Cu = $CuSO_4$, Fe = $FeSO_4$, Sn = $SnCl_2$

Table 10. Colour fastness to water (ISO 105-E01: 1994)

Fastness	Pad-batch					Pad-dry				
	Without	Al	Cu	Fe	Sn	Without	Al	Cu	Fe	Sn
Acid										
Colour change	4-5	4-5	4-5	4-5	4-5	4-5	4-5	4-5	4-5	4-5
Colour staining										
-Acetate	4-5	4-5	4-5	4-5	4-5	4-5	4-5	4-5	4-5	4-5
-Cotton	4	4-5	4	4	4-5	4	4-5	4	4	4-5
-Nylon	4-5	4-5	4-5	4	4-5	4-5	4-5	4-5	4	4-5
-Polyester	4-5	4-5	4	4	4-5	4-5	4-5	4	4	4-5
-Acrylic	4-5	4-5	4-5	4-5	4-5	4-5	4-5	4-5	4-5	4-5
-Wool	4-5	4-5	4-5	4	4-5	4-5	4-5	4-5	4	4-5
Alkaline										
Colour change	4-5	4-5	4-5	4-5	4-5	4-5	4-5	4-5	4-5	4-5
Colour staining										
-Acetate	4-5	4-5	4-5	4-5	4-5	4-5	4-5	4-5	4-5	4-5
-Cotton	4	4-5	4	4	4-5	4	4-5	4	4	4-5
-Nylon	4-5	4-5	4	4	4-5	4-5	4-5	4-5	4	4-5
-Polyester	4-5	4-5	4	4	4-5	4-5	4-5	4-5	4-5	4-5
-Acrylic	4-5	4-5	4-5	4-5	4-5	4-5	4-5	4-5	4-5	4-5
-Wool	4-5	4-5	4-5	4	4	4-5	4-5	4-5	4	4-5

Note: Al = $AlK(SO_4)_2$, Cu = $CuSO_4$, Fe = $FeSO_4$, Sn = $SnCl_2$

Table 11. Colour fastness to perspiration (ISO 105-E04: 1994)

7. Potential of eucalyptus leaves dye

7.1 Potential commercial applications

Natural dyes cannot be used as simple alternatives to synthetic dyes and pigments. They do, however, have the potential for application, in specified areas, to reduce the consumption of some of the more highly polluting synthetic dyes. They also have the potential to replace some of the toxic, sensitizing and carcinogenic dyes and intermediates (Deo & Desai, 1999). Eucalyptus leaves, as natural dye, has greater potential because it is grown already on an industrial scale. It also shows good fastness on wool substrate.

7.2 Potential effluent problems

The effluent problems of synthetic dyes occur not only during their application in the textile industry, but also during their manufacture, and possibly during the synthesis of their intermediates and other raw materials. The application of synthetic dyes also requires metal salts for exhaustion, fixation, etc (Deo & Desai, 1999). Natural dyes, like eucalyptus leaves do not cause damage the environment by their extraction and many could be used satisfactorily without mordants, although it is true that the use of mordant improves the depth of shade for natural dyes. These mordants are normally metal salts and hence damage to the environment is still possible, albeit to a smaller extent than for synthetic dyes in textile applications. The research in this field has already identified a few "natural mordant", such as *Entada spiralis Ridl* (Chairat et al., 2007) and harda (*Chebulic myrabolan*) (Deo & Desai, 1999). The avoidance of metal-based mordants, or their replacement by natural mordants, may assist in the preservation of the environment.

8. Acknowledgement

The authors would like to thank Ing. Martina Viková from the Technical University of Liberec, Czech Republic, for testing of the UV transmission and UPF values.

9. Conclusions

A wool fabric dyed in a solution containing the eucalyptus leaf extract showed a shade of pale yellowish-brown. The exception was when the fabric was dyed with the ferrous sulfate mordant, resulting in a shade of dark greyish-brown. The yield (exploitation) of the coloring component of eucalyptus leaf extract is good in wool fabric (about 68%–52% from the lowest to the highest concentrations). It can be observed that the K/S values increase with an increase of dye concentration. Little difference between the two padding techniques utilized for the wool fabric dyes by three mordanting methods, except wool fabrics mordanted with copper sulfate whose gave a high K/S values on the pad-batch technique than pad-dry technique. In all cases ferrous sulfate mordant yielded the best dyeing results, and the next good result was obtained in the order of copper sulfate, stannous chloride and alum. As observed from the K/S values, in the case of wool fabrics dyed with alum by using post-mordanting method gave lower colour strength than without mordant. Alum and ferrous sulfate were the best mordant during simultaneous mordanting method of dyeing. However, copper sulfate showed the best mordant during simultaneous mordanting and pre-mordanting method of dyeing. For the K/S value on dyed wool fabrics were only little different using stannous chloride as mordant during three mordanting methods. The fastness properties ranged from good to excellent, while light fastness was fair to good. It was observed that the ultraviolet (UV) protection factor (UPF) values rated as excellent for the wool fabric. In addition, a darker colour, such as that provided by a ferrous sulfate mordant, gave better protection because of higher UV absorption.

The application of eucalyptus leaves dye on wool fabrics by pad-batch and pad-dry technique of dyeing can be considered as an affective eco-option because it gives extremely good results with substantial minimization of processing cost. In case of pad-dry technique, the average hot air consumption is considerably high whereas no hot air is being consumed in cold pad-batch process which leads to energy conservation. However, the time employed for the fixation of eucalyptus leaves dye is very long in cold pad-batch technique. So, these techniques can be considered as best suitable for small scale industries or cottage dyeing of eucalyptus leave.

10. References

Allen, R. L. M. (1971). *Colour chemistry*, Nelson, ISBN 01-77617-17-9, London, England

Samanta, A. K. & Agarwal, P. (2009). Application of natural dyes on textiles. *Indian Journal of Fibre & Textile Research*, Vol. 34, No. 4, (December 2009), pp. 384-399, ISSN 0971-0426

Bechtold, T. & Mussak, R. (2009). *Handbook of natural colorants*, John Wiley & Sons, ISBN 978-0-470-51199-2, West Sussex, England

Vankar, P. S. (2007). *Handbook on natural dyes for industrial applications*, National Institute of Industrial Research, ISBN 81-89579-01-0 ,New Delhi, India

Glover, B. & Pierce, J. H. (1993). Are natural colorants good for your health? *Journal of the Society of Dyers and Colourists*, Vol. 109, No. 1, pp. 5-7, ISSN 0037-9859

Hwang, E. K. ; Lee, Y. H. & Kim, H. D. (2008). Dyeing, fastness, and deodorizing properties of cotton, silk, and wool fabrics dyed with gardenia, coffee sludge, *Cassia tora. L.*, and pomegranate extracts. *Fibers and Polymers*, Vol. 9, No. 3, pp. 334-340, ISSN 1857-0052

Hunger, K. (2003). *Industrial dyes: Chemistry, Properties, Applications*, WILEY-VCH Verlag GmbH & Co. KGaA, ISBN 35-27304-26-6, Darmstadt, Germany

Flint, I. (2007). An antipodean alchemy–The eucalypt dyes. *Turkey Red Journal*, Vol. 13, No. 1, (Fall 2007). Available from : http://www.turkeyredjournal.com/archives/V13_I1/flint.html

Ali, S. ; Nisar, N. & Hussain, T. (2007). Dyeing properties of natural dyes extracted from eucalyptus. *The Journal of The Textile Institute*, Vol. 98, No. 6, pp. 559-562, ISSN 1754-2340

Vankar, P. S. ; Tiwari, V. & Srivastava, J. (2006). Extracts of steam bark of *Eucalyptus Globules* as food dye with high antioxidant properties. *Electronic Journal of Environmental, Agricultural and Food Chemistry*, Vol. 5, No. 6, pp. 1664-1669, ISSN 1579-4377

Chapuis-Lardy, L. ; Contour-Ansel, D. & Bernhard-Reversat, F. (2002). High performance liquid chromatography of water-soluble phenolics in leaf litter of three eucalyptus hybrids (Congo). *Plant Science – Kidlington*, Vol. 163, No. 2, pp. 217-222, ISSN 0168-9452

Conde, E.; Cadahia, E. & Garcia-Vallejo, M. C. (1997). Low molecular weight polyphenols in leaves of *Eucalyptus camaldulensis. E. globules and E. rudis. Phytochemical Analysis*, Vol. 8, No. 4, pp. 186-193, ISSN 0958-0344

Conde, E.; Cadahia, E.; Garcia-Vallejo, M. C. & Fernandez de Simon, B. (1997). High pressure liquid chromatographic analysis of polyphenols in leaves of *Eucalyptus cmadulensis. E. globules* and *E. rudis*: proanthocyanidins, ellagitannins and flavonol glycosides. *Phytochemical Analysis*, Vol. 8, No. 2, pp. 78-83, ISSN 0958-0344

Mongkholrattanasit, R. ; Wongphakdee, W. & Sirikasemlert, C. (2007). Dyeing and colour fastness properties of silk and cotton fabrics dyed with eucalyptus bark extract. *RMUTP Research Journal*, Vol. 1, No. 1, pp. 41-49, ISSN 1906-0432

Mongkholrattanasit, R. ; Kryštůfek, J. & Wiener, J. (2010). Dyeing and fastness properties of natural dye extracted from eucalyptus leaves using padding techniques. *Fibers and Polymers*, Vol. 11, No.3, pp. 346-350. ISSN 1875-0052

Mongkholrattanasit, R. ; Kryštůfek, J. ; Wiener, J. & Viková, M. (2011). Dyeing, fastness, and UV protection properties of silk and wool fabrics dyed with eucalyptus leaf extract by exhaustion process. *FIBRES and TEXTILES in Eastern Europe Journal*, Vol. 19, No. 3, pp.94-99, ISSN 1230-3666.

Vankar, P. S. ; Tivari, V. ; Singh, L. W. & Potsangbam, L. (2009). Sonicator dyeing of cotton fabric and chemical characterization of the colorant from *Melastoma malabathricum*. *Pigment & Resin Technology*, Vol. 38, No. 1, pp. 38-42, ISSN 0369-9420

Tiwari, V. & Vankar, P. S. (2001). Unconventional natural dyeing using microwave and sonicator with alkanet root bork. *Colourage*, Vol. 25, No. 5, pp. 15-20, ISSN 0010-1826

Moeyes, M. (1993). *Natural dyeing in Thailand*, White Lotus, ISBN 974-8495-92-2, Bangkok, Thailand

Mongkholrattanasit, R. ; Kryštůfek, J. & Wiener, J. (2009). Dyeing of wool and silk by eucalyptus leaves extract. *Journal of Natural Fibers*, Vol. 6, No. 4, pp. 319-330, ISSN 1544-0478

Agarwal, B. J. & Patel, B. H. (2002). Studies on dyeing of wool with a natural dye using padding techniques. *Man-Made Textiles in India*, Vol. 45, pp. 237-241, ISSN 0377-7537

Yarosh, E. A. ; Gigoshvili, T. I. & Alaniya, M. D. (2001). Chemical composition of *Eucalyptus jumanii* cultivated in the humid Georgian subtropics. *Chemistry of Natural Compound*, Vol. 37, No. 1-2, pp. 86-87, ISSN 1573-8388

Pretsch, E.; Bühlmann, P. & Affolter, C. (2000). *Structure determination of organic compounds (Tables of spectral data)*, Springer-Verlag, ISBN 978-3-540-67815-1, Berlin, Germany

Feng, X. X. ; Zhang, L. L ; Chen, J. Y. ; & Zhang, J. C. (2007). New insights into solar UV-protectives of natural dye. *Journal of Cleaner Production*, Vol. 15, No.4, pp. 366-372, ISSN 0959-6526

Sarkar, A. K. & Seal, C. M. (2003). Color strength and colorfastness of flax fabrics dyed with natural colorants. *Clothing and Textiles Research Journal*, Vol. 21, No. 4, pp. 162-166, ISSN 1940-2473

Giles, C. H. (1974). *A Laboratory course in dyeing*. Society of Dyers and Colourists, ISBN 1940-2473, Yorkshire, England

Duff, D. G. & Sinclair, R. S. (1989). *Giles's Laboratory course in dyeing*. Society of Dyers and Colourists, ISBN 09-01956-49-X, West Yorkshire, England

Bhattacharya, S. D. & Shah, A. K. (2000). Metal ion effect on dyeing of wool fabric with catechu. *Coloration Technology*. Vol. 116 , No. 1, pp. 10-12, ISSN 1472-3581

Gies, P.H.; Roy, C.R. & Holmes, G. (2000). Ultraviolet radiation protection by clothing: Comparison of in vivo and in vitro measurements. *Radiation Protection Dosimetry*, Vol. 91, No.1-3, pp. 247-250, ISSN 0144-8420

Allen, M.W. & Bain, G. (November 1994). *Measuring the UV protection factor of fabric*. Retrieved March 25, 2008, from Varian Australia Pty Ltd. Available from : http://www.thermo.com/eThermo/CMA/PDFs/Articles/articlesFile_6716.pdf

Deo, H. T. & Desai, B. K. (1999). Dyeing of cotton and jute with tea as a natural dye. *Journal of the Society of Dyers and Colourists*. Vol. 115, No. 7-8, pp. 224-227, ISSN 0037-9859

Sarkar, A. K. (2004). An evaluation of UV protection imparted by cotton fabric dyed with natural colorants. *BMC Dermatology*, Vol. 4 , No.15, pp. 1-8, ISSN 14715945

Chairat, M. ; Bremner, J. B. & Chantrapromma, K. (2007). Dyeing of cotton and silk yarn with the extracted dye from the fruit hulls of mangosteen, *Garcinia mangostana* Linn. *Fibers and Polymers*, Vol. 8, No. 6, pp. 613-619, ISSN 1875-0052

6

Flame Retardancy and Dyeing Fastness of Flame Retardant Polyester Fibers

Seung Cheol Yang, Moo Song Kim and Maeng-Sok Kim

Nylon Polyester Polymer R&D Team, Production R&D Center, Hyosung R&D Business Labs, 183, Hogye-dong, Dongan-gu, Anyang-si, Korea

1. Introduction

This is a continuation of previously published paper[1, 2, 3] in which the dyeing properties and chemical resistances of the flame retardant polyester fiber according to the type of flame retardants were reported.

Polyester, mainly poly(ethylene terephthalate) fiber is widely used for textile apparel, industrial fiber, tire cord, etc., and the demands and supplies are growing annually by 9%. As interest in the danger of fire, the demand for flame retardant polyester fiber has been strong and there have been many researches and developments to improve the flame retardant polyester.

In the our previous publication[1, 2, 3], we report the relationships between phosphorous flame retardants and the products such as polymer, fiber according to the type of flame retardants. Flame retardant polyester fiber is used for other purposes not as a normal PET fiber. Normal PET fiber is primarily used in the textile apparel whereas the flame retardant polyester fiber is mainly used for industrial applications such as the upholstery or car interior. Therefore, flame retardant polyester fiber goods are exposed to the open air more than a normal PET fiber. If the flame retardant polyester fiber vulnerable to UV, the application of the fiber is limited. So, the flame retardant polyester fiber must have good weatherability such as good light fastness and anti-UV properties.

In general, Polyester fiber use deluster to reduce the transmission of light. Deluster, commonly titanium dioxide, shows good protection of UV in the fields, preventing the passage of light. However, the titanium dioxide has low band gap, between the ground state and the excited state, it is deteriorated in high dosage and for long periods.[4]

Dyeing of polyester products primarily conducted at high temperature and pressure using disperse dyestuff. Disperse dyestuff is divided into nitrodiphenyl, amine, azo and anthraquinone dyestuff greatly depending on its chemical structure.[5]

In this chapter, flame retardancy and dyeing fastness of flame retardant polyester fiber were compared with those of normal PET fiber in accordance with contents of the flame retardant and deluster. Increasing deluster(titanium dioxide) content in the flame retardant polyester fiber, flame retardancy is lowered, but dyeing fastness shows almost same level. Besides light fastness, dyeing fastness shows similar level with increased phosphorous content compared with normal polyester fiber.

2. Experiments & results

2.1. Fiber preparation

Flame retardant polyester polymers using phosphorous flame retardant were prepared in a same method as the previous report [2]. 3-(Hydroxyphenyl phosphinyl) propanoic acid was used as a phosphorous flame retardant, and was supplied from I company in Korea and used without further purification. The phosphorous content was adjusted 0.65wt% as a phosphorous atom on the basis of flame retardant polyester, and the contents of titanium dioxide were adjusted 0, 0.3, 2.5wt% respectively. Generally, the polyester fibers were called the as bright, semi dull and full dull according to the contents of deluster(titanium dioxide). The brief chemical formula of the flame retardant polyester is shown in Figure 1.[2]

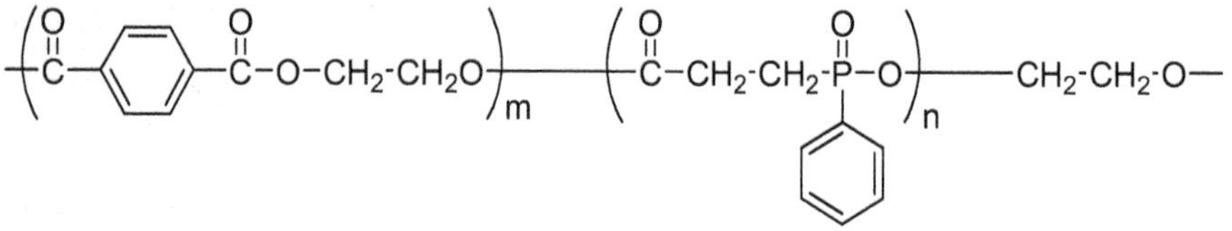

$$\left(\!\!-\overset{O}{\overset{\|}{C}}-\!\!\!\bigcirc\!\!\!-\overset{O}{\overset{\|}{C}}-O-CH_2\text{-}CH_2O\!\!-\!\!\right)_{\!\!m}\ \ \left(\!\!-\overset{O}{\overset{\|}{C}}-CH_2\text{-}CH_2\text{-}\overset{O}{\overset{\|}{P}}-O\!\!-\!\!\right)_{\!\!n}\ \ -CH_2\text{-}CH_2\text{-}O-$$

Fig. 1. Chemical formula of the flame retardant polyester

The flame retardant polyester polymers were dried in vacuo, the humidity was less than 25ppm based on polymer weight. And as previously reported, the 75d/36f filament yarns were prepared by spin-draw method.

Dyeing were conducted using hose-knitted sample and under the same condition as previously report. Commercially available anthraquinone disperse dyestuff was used for the dyeing test.

Flame retardancies and Dyeing fastnesses were evaluated by FITI Testing & Research Institute in Korea on the standard of LOI(limit oxygen index, KS M ISO 4859-2:2001) and KS K 0903, respectively.

2.2 Flame retardancy with deluster contents

LOIs according to the titanium content in the flame retardant polyester fiber were shown in Figure 2.

LOI is linearly decreased along with increase of titanium dioxide contents.

Decrease of flame retardancy due to titanium dioxide is not clearly defined, however, it is assumed that by phtocatalytic property of anatase type of titanium dioxide. Titanium dioxide used in this experiment was anatase type produced form Sachtleben Company in Germany under the brand name of Hombitan LWSU. In polyester fiber industry, anastase type titanium dioxide is usually adopted. Another type of titanium dioxide, rutile type, is not adopted in polyester fiber industry. Between both types of titanium dioxides(anatase vs rutile) , anatase type titanium dioxide has a bandgap of 3.2eV [6,7] as shown in Figure 3.

In general, most of the phosphorous compounds, including phosphorous flame retardants in this chapter, are known to lower the photocatalytic activity of titanium dioxide. It is assumed that the flammability is decreased by increase amount of titanium dioxide which has photocatalytic activity[8]. However, if the good has the LOI value over 30, it is known that it is good flame retardant performance.

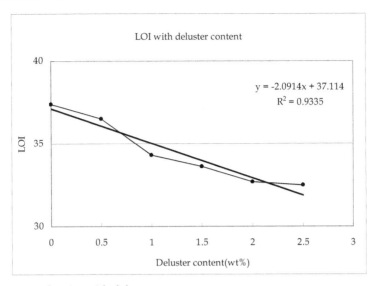

Fig. 2. Flame retardancies with deluster contents

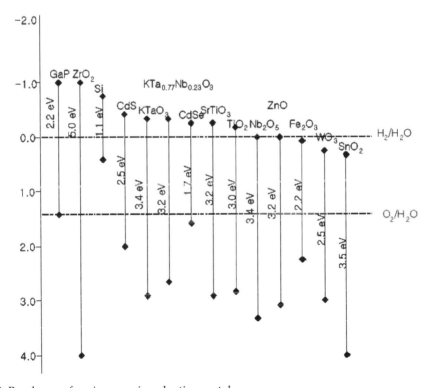

Fig. 3. Bandgaps of various semiconducting metals

2.3 Dyeing fastnesses with deluster contents

Table 1 shows dyeing fastness of the flame retardant polyester fibers along with the contents of titanium dioxide used as Deluster.

Deluster conent(wt%)	Washing	Rubbing	Sublimation	Light
0%	4~5	4~5	4~5	3
0.3%	4~5	4~5	4~5	3
2.5%	4~5	4~5	4~5	3
Normal PET(0.3%)	3~4	4~5	4~5	5

Table 1. Dyeing fastnesses(grade) with deluster contents.

Among dyeing fastnesses, rubbing and sublimation fastness showed similar trend with Normal PET, but washing fastness was better than Normal PET, light fastness was lower than Normal PET.

It is assumed that washing fastness is related to chain mobility of polyester polymer caused by the difference of glass transition temperature. Flame retardant polyester fiber has lower glass transition temperature than normal PET. Therefore, it has better chain mobility than normal PET, and the uptake of the dyestuff at the same temperature is higher than that of normal PET. 2)

Decrease of light fastness is assumed that the influence of photocatalytic activity of titanium dioxide above mentioned. Under the same condition, light fastness using azo disperse dyestuff has 3~4 grade of level which is somewhat better than that using anthraquinone disperse dyestuff, but the dyeing fastness of the flame retardant polyester fiber shows still inferior property compared to normal PET.

2.4 Light fastnesses with phosphorous contents

We can see the dyeing fastnesses are almost same level of flame retardant polyester and normal polyester except for the light fastness as shown in Table 1.

The effect of phosphorous content in the flame retardant polyester on Light fastness was conducted by varying the phosphorous contents in the polymer(deluster contents are same as 0.3% by weight), and the result is shown in Table 2.

Phosphorous content (wt%)	0(Normal PET)	0.5%	0.65%
Light fastness	5	3~4	3

Table 2. Light fastness with phosphorous contents.

Light fastness of Flame retardant polyester fiber is inferior to that of normal PET. The difference is brought about by the content of phosphorous flame retardant, not by the content of titanium dioxide. For that reason, there is a necessity for flame retardant polyester fiber to improve light fastness for using industrial applications such as upholstery and car interior, etc.

2.5 Light fastnesses improvement

To improve the light fastness of the flame retardant polyester fiber, following method was chosen and conducted ;

1. Increase of inorganic UV stabilizer(manganese acetate) in polymer

Manganese acetate is known to be incorporated into the polymer as UV stabilizers under polymerization of flame retardant polyester[9]. Flame retardant polyester fibers incorporated manganese acetate as the basis of 60, 120, 200, 300 ppm of manganese metal in polymer were polymerized. And the polymers were spun in a same method as above mentioned. The results of the light fastness were shown in Table 3.

Manganese content(ppm)	60	120	200	300
Light fastness	3~4	3	3	3

Table 3. Light fastness with manganese contents.

Table 3 shows that the light fastness decreases with the content of manganese metal. Above all, the coagulation of the manganese acetate brought about the increase of filtration pressure in the polymerization and spinning process to conducting long period test. If the light fastness is improved, this method would be difficult to apply to commercial production of flame retardant polyester fiber.

2. Incorporation of organic UV stabilizer into polymer

To improve the light fastness of the flame retardant polyester fiber, some UV stabilizers were recommended by UV stabilizer maker. The basic properties of the UV stabilizers used in this test were shown in Table 4. The UV stabilizers were incorporated in polymerization reactor as in the state of ethylene glycol dope and the contents of the UV stabilizers were adjusted to 0.3 wt% based on the polyester polymer. The fibers are prepared as same method above mentioned.

Light fastnesses of all flame retardant polyester fibers incorporated various UV stabilizers how negligibly little difference as compared with that not containing the UV stabilizers.(same grade of light fastness of 3)

3. Using of UV absorber under dyeing

Benzotriazole type UV absorber as dyeing auxiliary was recommended form dyestuff maker and dyeing test was conducted. It is added to dyeing solution whose concentration is owf 1%, and the other conditions are same as the above dyeing method. The light fastnesses are increased to the grade over 4(at least 1 grade better that of not treated good).

To investigate the effect of benzotriazol UV absorber on the light fastness of polyester fibers, the UV absorption spectrum is shown in Figure 4.

UV Absorber	Chemical structure	Molecular weight	Melting Point (°C)	Remark
Chimassorb 119FL		2286	115~150	Triazine derivative
Flamestab NOR 116		2261	108~123	Triazine derivative
Hostavin ARO 8P		326	>48	Benzophenone derivative
Hostalux KS			250~300	Benzoxazole derivative

Table 4. UV stabilizers used in the test

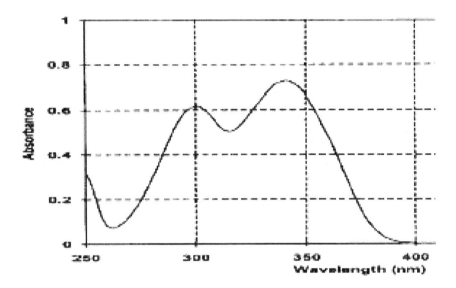

Fig. 4. UV absorption spectrum of benzotrazole UV absorber

Benzotriazol UV absorber is thought to improve the light fastness of the polyester fiber by absorbing the light around 350nm wavelength which is mainly absorbed in polyester. Improving the light fastness using benzotriazol UV absorber can be adopted not only to flame retardant polyester fibers but also to the normal PET fibers.

4.   Comparison of the disperse dyestuff

It would be the best way to find a suitable dyestuff to improve light fastness of the flame retardant polyester fiber if it is commercially available. The effect of the disperse dyestuff on light fastness of the flame retardant polyester fiber was compared using the dyestuff kind according to their chemical structure.

From the standpoint of chemical structure, the disperse dyestuffs were categorized into nitrodiphenyl amine dyestuff, azo dyestuff and anthraquinone dyestuff. Among these dyetuffs, the azo and anthraquinone dyestuffs are mainly used in commercial scale.

Dyestuffs	Anthraquinone dyestuff	Azo dyestuff (Red)	Azo dyestuff (Blue)
Light fastness	3	Above 4	4

Table 5. Light fastness according to the dyestuff

In Table 5, there is a noticeable difference in light fastness dyed with between azo dyestuff and anthraquinone dyestuff. The light fastness dyed with anthraqunone dyestuff is much inferior to that with azo dyestuff.

It is assumed that insoluble dyestuff is reduced to leuco compound by the high reductivity of the phenyl phosphinic acid in the phosphorous flame retardant [2, 6] as shown in Figure 5.

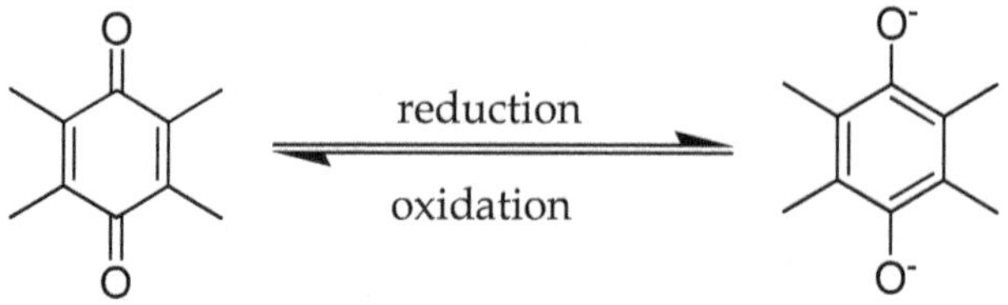

Fig. 5. The brief scheme of reduction state of anthraquinone dyestuff

3. Conclusion

In this chapter, the flame retardancies with deluster contents and dyeing fastnesses with phosphorous contents were investigated.

Whereas titanium dioxide as the deluster affects the flame retardancy of the polyester, it does not affect the dyeing fastness. Titanium dioxide is semi conductor material whose band gap is 3.2eV. Titanium compound in polyester fiber is activated by light, so polyester fibers containing titanium dioxide has color shading problem(yellowing) for outdoor usage. To minimize the yellowing of the polyester goods of polyester fiber with titanium dioxide, phosphorous stabilizer was used. So it could be thought that a part of phosphorous compound act as the stabilizer to reduce the activity of titanium dioxide and the remnant reveals the flame retardancy.

Low level of light fastness of phosphorous flame retardant polyester fiber is due to reductivity of the phosphorous compound. The phosphorous compound shows acidity by the nature of phenyl phosphinic acid to fade out the disperse dye. Light fastness of the phosphorous flame retardant polyester can be minimized by benzotriazole stabilizer[5]. In commercial scale, it is a proper way to improve of light fastness of Flame retardant polyester fiber.

4. References

[1] S.C.Yang; J.P.Kim(2007). *J. Appl. Polym. Sci.*, Vol.106, No.5, pp.2870-2874
[2] S.C.Yang; J.P.Kim(2007). *J. Appl. Polym. Sci.*, Vol.108, No.2, pp.1274-1280
[3] S.C.Yang; J.P.Kim(2008). *J. Appl. Polym. Sci.*, Vol.108, No.4, pp.2297-2300
[4] M.Gratzel(1997), *Studies in Surface Science and Catalysis*, Vol.103, pp.353-375
[5] D.M.Nunn, *The dyeing of synthetic-polymer and acetate fibres*, pp.10~12, London, Dyers Company Publications Trust(1979)
[6] K. Madhusudan Reddy et al.(2002), *Materials Chemistry and Physics*, Vol.78, pp. 239–245
[8] L. K″or″osi, I. D´ek´any (2006), *Colloids and Surfaces A: Physicochem. Eng. Aspects*, Vol. 280, pp. 146–154
[9] S.C.Yang, et al., WO 2006/070969
[10] S.G.Park, *in Seminar Handout*, 13th. Nov. 2001.

Part 3

Surface Modification

7

Protein Fibre Surface Modification

Jolon Dyer[1,2,3,4] and Anita Grosvenor[1]
[1]AgResearch Lincoln Research Centre, Christchurch,
[2]Lincoln University, Canterbury,
[3]Biomolecular Interaction Centre, University of Canterbury, Christchurch,
[4]Riddet Institute at Massey University, Palmerston North,
New Zealand

1. Introduction

Many natural fibres, including wool, cashmere and silk, are protein-based materials; the dry weight of wool is almost entirely derived from proteins (Maclaren & Milligan 1981). As such, they possess an inherent structural and chemical heterogeneity not found in synthetic polymers. Although typically less heterogeneous than biological fibres, the rapidly emerging range of commercially available protein-based biomaterials also contain a wide range of functionality derived from their constituent primary and secondary protein structure.

The response of fibres to processes such as dyeing and finishing treatments correlates directly to their structural and chemical properties, and this is particularly true for surface treatments. Due to its barrier function in the fibre, modification of the surface has a profound impact on processing and performance. Keratinous fibres such as wool and cashmere have an outer lipid layer which results in a hydrophobic surface. Recently a range of innovative and novel fibre surface technologies has been developed, many of which involve altering surface properties by the removal of the lipid layer, which exposes a proteinaceous surface with a variety of reactive chemical moieties. Treatments that can be covalently bound to fibre surface components, rather than simply physically applied to the surface, offer the potential for superior durability.

The surface modification of proteinaceous fibres has a long history. These include plasma applications, which expose and generate functional groups on the protein surface, etching into the surface of the cuticle scales, to improve properties such as surface wettability, dyeability, shrink-resistance and felting-resistance. Chemical approaches utilised include ozone treatments, which cause oxidation of the surface and altered ionic balance, leading to a more plastic and reactive fibre surface and shrinkage control; chlorination, which improves sorption characteristics and reduces shrinkage; hydrogen peroxide treatment; and acid anhydride acetylation of silk and wool, which improves the textile response to dyeing, shrink resist, and setting treatments. Enzymatic treatments have been utilised to decuticulate the surface and improve properties such as shrink resistance. More recent developments include reaction of functional chemical agents or branched molecules to exposed reactive groups on the fibre surface, enabling the attachment of covalently-bound smart treatments, or the amplification of reactive groups for increased functionality.

This chapter outlines developments in the area of targeted surface modification of protein-based fibres and textiles, including summarising applications and future directions. It is not intended to cover every aspect of fibre surface modification, but rather focus on key types of surface modification that exploit the unique properties of proteinaceous fibres.

2. Physical modification

The increasing pressure for environmentally friendly processing approaches has motivated the textile industry to exchange aqueous treatments involving potential chemical pollutants and effluents with high adsorbable organic halide (AOX) contents for physical 'dry' processes, such as plasma and UV/ozone treatment. With respect to dyeing, the nature of physical modification to protein fibre surfaces and where this occurs in production are important and must be taken into consideration.

2.1 Plasma

Plasma treatments, which utilise a gaseous electrical discharge, are reported to be surface specific for protein fibres (Höcker 2002) and offer significant potential in terms of being simple, clean, solvent-free and relatively inexpensive. These treatments can be used to modify surfaces by the deposition of polymers or can 'clean' the surface by surface etching. Plasma treatments are increasingly replacing wet (chemical) textile treatments to achieve outcomes such as shrink-resistance and improved dyeability. Plasma and corona treatments oxidise the surface of protein-based textiles, generate chemically active radicals, induce functionalisation, and etch the surface (Höcker 2002; Ceria et al. 2010). Those utilising certain gases may result in the deposition of atoms from those gases, such as fluorine (Höcker 2002). The effects of plasma treatments are restricted to the wool surface, and are therefore unlikely to result in changes to the bulk properties resulting from damage to the interior of the fibres.

Within the context of the textiles industry, plasma means the products of the interaction of an electromagnetic field with gas; namely, a partially ionised gas that contains ions, electrons and neutral particles (Kan & Yuen 2007). At gas pressures similar to atmospheric pressure and high voltage, corona discharge is generated. At gas pressures of 0.1-10 MPa at lower voltage, glow discharge is generated (Rakowski 1997; Kan & Yuen 2007). Atmospheric corona discharge plasmas lack uniformity due to their filamentous nature (Prat et al. 2000). Glow discharge is most commonly referred to when plasma textile processes are described. Many plasma treatments have employed low pressure (vacuum) systems to keep the plasma stable, called cold plasma or non-equilibrium plasma. Lowered pressure also assists with the penetration of plasma effects through the thickness of a textile (Poll et al. 2001). Atmospheric pressure plasma systems, however, have an industrial advantage for large-scale textile applications because of the expense, time and space involved in maintaining a vacuum (Sugiyama et al. 1998; Demir 2010). These require a source frequency of 1-20 kHz and a carrier gas of helium (Prat et al. 2000). Microwave-induced glow-discharge plasma can be kept stable at atmospheric pressure (Sugiyama et al. 1998). Several types of discharge may be used to generate plasma, including direct current discharge, radio-frequency discharge, and microwave.

In wool, plasma treatment oxidises and partially removes (ablates) the hydrophobic surface lipid layer (both the loosely adhering lipids and those covalently bound). The disulfide bonds in the surface protein layer of wool (epicuticle) are also oxidised (Höcker 2002). The surface is the only part of the fibre affected; this becomes rougher (surface area increases), while its protein contents are hardly affected (Höcker 2002). The free radicals that remain on the wool surface following ablative plasma treatment stimulate the formation of functional groups and of bonds between the fibre surface and customised surface coatings (Kan &

Yuen 2007). As an alternative to ablation, with the use of carbon- and hydrogen-rich plasma gases, polymers may be deposited on the surface (Kan & Yuen 2007).

The physical changes (roughening) caused by ablative plasma treatments result in altered yarn strength, frictional and spinning properties and decreased felting behaviours (Höcker 2002). A detrimental effect is a firm or harsh handle (Rakowski 1997). The chemical changes (surface oxidation and the loss of surface-bound fatty acids) result in improved dyeing and resistance to shrinkage (Rakowski 1997). Shrink-resistance is obtained at levels comparable to those obtained from other treatments (like chlorination, reduction and resin application) by the addition of resin, causing surface smoothing. Dyeing depth, speed and bath exhaustion are also improved (Höcker 2002). Atmospheric pressure plasma treatment is also seen to improve the dyeability of mohair (Demir 2010). An unusual application of wool plasma treatment is the deposition of antibacterial polymers for woollen wound dressings using fluorinated post-discharge plasmas (Canal et al. 2009).

Due to the wide variety of techniques utilised to treat protein-based fibres with plasma, it can be difficult to compare the effects reported by different groups. Relating this problem to dyeing, Naebe et al. showed that the effect of plasma on dyeing depended on the dye used – with hydrophobic dyes unaffected by plasma treatment, and hydrophilic dyes greatly affected (Naebe et al. 2010). One general feature of plasma-based approaches to modifying the surface of protein materials is the high level of surface oxidation observed after treatment (Meade et al. 2008b).

Hydrophilicity can be a desirable attribute for textiles, allowing improved dyeing, and enhancing comfort and wear properties. The natural hydrophobic surface of wool makes it practically impossible to print wool fabrics without prior treatment. Plasma treatment improves fabric wettability, and also reduces felting (Kan & Yuen 2007). This can be accomplished by the removal of surface lipids (oxygen plasma) or by the deposition of hydrophilic monomers, such as the plasma-assisted deposition of acrylic acid (Kutlu et al. 2010). Aside from dye properties and processing, the wool dyeing rate is primarily determined by fibre morphology and by the state of adsorbed water in the fibre (Ristić et al. 2010). These properties may be modified by a variety of surface treatments. When the outer lipid layer is even partially removed from wool, the wettability and dyeing rate are reduced. Plasma treatments remove this lipid layer and generate functional groups (such as thiols) that are more reactive to certain dyes. Corona discharge has been reported to incorporate oxygen atoms into the fibre and enhances wettability, which increases the acid-dye intensity of printed fabrics (Ryu et al. 1991). Combining plasma treatments with chitosan results in increases in colour intensity and dyeability over those observed with either treatment alone (Ristić et al. 2010).

2.2 UV/Ozone

UV/ozone (UVO) treatments also achieve textile improvement effects able to compete with aqueous treatments. UV radiation at certain frequencies generates atomic oxygen and ozone from molecular oxygen (at 184.9 nm) and atomic oxygen from ozone (at 253.7 nm). Organic hydrocarbons may also be excited at 253.7 nm. Textiles placed close (within 5 mm) of these UVO sources are cleaned by the reaction of UV-excited surface molecules with atomic oxygen. The volatile reaction products desorb from the surface (Shao et al. 1997).

This surface treatment results in more wettable wool, which improves dyeing and printing properties, even at low temperature (Xin et al. 2002). The wool also yellows, although this can be lessened when the process is combined with peroxide pad bleaching (Shao et al.

2001). This is due to oxidation of the surface layers: Cysteic acid is detected, as disulfide bonds are broken. Carboxylic acid and carbonyl groups are detected at higher levels than after chlorination (Shao et al. 1997). The lipids at the surface are modified or volatised. UVO treatment results in a colour yield and dyeability comparable to that obtained after chlorination, permitting printing (Shao et al. 1997; 2001).

3. Chemical modification

3.1 Bleaching

Wool bleaching affects the fibre surface, but is not performed specifically to produce functionalised surfaces, so will not be discussed at length in this chapter. Agents utilised in protein-fibre bleaching include hydrogen peroxide, sodium borohydride, sodium bisulfite, thiourea, oxalic acid and blue light (Arifoglu et al. 1992; Millington 2005; Yilmazer & Kanik 2009). Bleaching is sometimes combined with additional surface modification techniques such as scale removal (Chen et al. 2001).

3.2 Acylation

Acylation may provide a route to water-proofing of materials such as wool and silk. Acylation agents include dodecenylsuccinic anhydride and ctadecenylsuccinic anhydride, anhydrides such as succinic anhydride and phthalic anhydride, and solvents including dimethyl sulfoxide and dimethylformamide (Arai et al. 2001; Davarpanah et al. 2009).

Wool is noted to gain more weight and acyl content than silk. Silk tensile properties are unaffected, whereas wool displays greater extensibility. Silk and wool both increase in water repellency and decrease in moisture regain and water retention. Deposits are observed (via SEM), which are attributed to the modifying agents (Arai et al. 2001). Acylation provides an enhanced surface for the grafting of chitosan to wool or silk to provide antibacterial and anti-felting properties and superior dyeing ability in an environmentally friendly fashion (Davarpanah et al. 2009; Ranjbar-Mohammadi et al. 2010).

3.3 Chlorination

Chlorination is performed to impart shrink-resistance to wool fibre, sometimes in combination with the applications of resins such as Hercosett or Nopcobond (Van Rensburg & Barkhuysen 1983; Roeper et al. 1984). As an industrial process, this is under regulatory pressure because of the AOX released into the waterways, raising environmental concerns.

Chlorination mixtures are generated by the combination of hypochlorite with sulfuric acid, or, for a milder treatment, chlorine gas may be dissolved in water (Van Rensburg & Barkhuysen 1983). Chlorination affects the surface lipid layer. Dyeing properties are also affected due to increased surface adsorption and alterations to the surface lipid layer (Ottmer et al. 1985; Baba et al. 2001). The changes in surface chemistry lead to more rapid dye exhaustion and dye penetration (Jocic et al. 1993).

3.4 Other

Potassium permanganate treatment of wool fibres reduces scale height and smoothes the cuticle, as assessed using 3D-SEM. This is desirable for the elimination of felting and to impart shrink resist. Potassium permanganate treatment provides a more even effect than a comparable proteolytic enzyme treatment (Bahi et al. 2007).

3.5 Delipidation

Keratin fibres such as wool, human hair and cashmere are covered in an outer lipid layer which is covalently bound to the surface to form a hydrophobic barrier. The major component of the surface wool lipids is 18-methyleicosanoic acid (18-MEA). 18-MEA is attached to the underlying protein mainly via covalent thioester bonding. A range of treatments have been reported to cleave the thioester bonds to form thiols on the epicuticle surface (Meade *et al.* 2008b). The generation of reactive surface sulfhydryl groups, with the sulfur able to act as a strong nucleophile, make these thiols attractive potential sites for subsequent covalent attachment of novel surface modifications (Meade *et al.* 2008a). Most research and development in this area has been performed with wool, but the principles have potential for application in other mammalian fibres used in textiles.

Wool fibres are comprised of a core of cortical cells surrounded by an outer sheath of overlapping cuticle cells. Each cuticle cell is enclosed within a resistant membrane termed the epicuticle (Höcker 2000). The epicuticle of wool covers the cuticle, and is comprised of both proteins and lipids (fatty acids); the hydrophobicity of the wool surface is largely attributable to this external lipid layer. The fatty acid component of the epicuticle accounts for approximately quarter of the epicuticle mass, with the surface-bound fatty acids forming a hydrophobic surface layer (Meade *et al.* 2008b). The branched chain fatty acid 18-methyleicosanoic acid (18-MEA) has been identified as the major lipid component of the wool surface, comprising approximately 65-70% of the surface lipid content (Negri *et al.* 1991; Ward *et al.* 1993). This 18-MEA is covalently bound to surface proteins via thioester linkages to cysteine, with the epicuticle estimated to have a content of 35% half-cystine (Negri *et al.* 1993; Evans & Lanczki 1997). The identity of the proteins that the surface lipids are bound to, forming proteolipids, is not yet well understood (Dauvermann-Gotsche *et al.* 1999). Thioesters are a relatively reactive group that can be cleaved relatively readily through nucleophilic substitution reactions.

If this outer lipid layer is removed in a controlled manner, it is possible to expose the underlying proteinaceous surface so that there are a variety of reactive functional groups (including hydroxyl, carboxyl and amine moieties) available for potential covalent attachment of surface treatments. There are various alkaline reagents that have been used to release covalently bound surface lipids from wool, yielding a hydrophilic and anionic surface with an increased frictional coefficient (Dauvermann-Gotsche *et al.* 2000).

The use of alcoholic or aqueous alkali conditions cleaves the thioester bonds to form a substituted lipid and a thiol at the epicuticle surface (Negri *et al.* 1991; 1993; Dauvermann-Gotsche *et al.* 2000) The chemical mechanism for nucleophilic substitution of the thioester bond is shown in Figure 1 [modified from Meade *et al.* 2008b].

Alcoholic alkali treatments use sodium butoxide, potassium tert butoxide, potassium hydroxide and hydroxylamine in water or in an anhydrous solvent such as tert butanol, dehydrated butanol, or ethanol (Leeder & Rippon 1983; Brack *et al.* 1996; Taki 1996; Meade *et al.* 2008b). These treatments are applied to control felting shrinkage, to improve polymer application, dyeing and printing, shrink resistance, and electrical conductivity (Leeder & Rippon 1983; Leeder *et al.* 1985). Alcoholic alkali treatment is less damaging than chlorination, as the effects are limited very much to the surface. The cortex is assumed to be unaffected; only the cuticle is removed, or de-scaled (Taki 1996). Lipid material is removed and a polar surface is generated. Covalently bound lipids are removed from the surface, revealing the polar surface of proteins beneath (Leeder & Rippon 1985; Brack *et al.* 1996; Meade *et al.* 2008b). As potassium t-butoxide is not found to be as good at removing the lipids as is the smaller potassium hydroxide, it is speculated that some lipids may be trapped within the proteins and that agent access plays a part in this difference. Removal of lipids may be made easier by prior fine scale damage (Brack *et al.* 1996). Functional agents revealed by removal of the lipid layer are available for covalent attachment of chosen treatments (Meade *et al.* 2008).

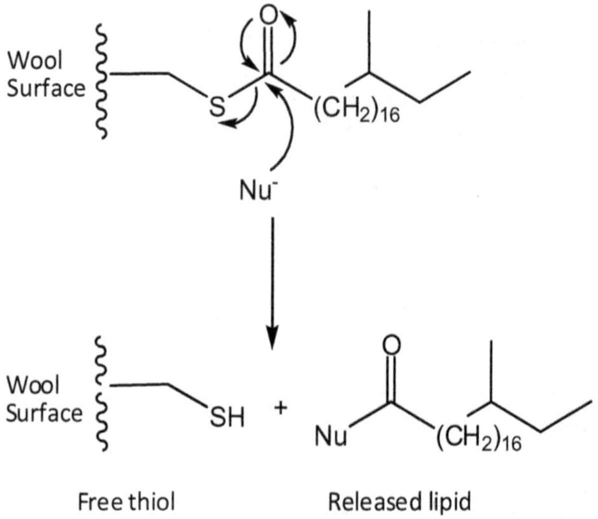

Free thiol   Released lipid

Fig. 1. Nucleophilic cleavage of 18-MEA from the wool protein surface (Meade *et al.* 2008b).

Methanolic potassium hydroxide is particularly effective at removing surface bound lipids from wool, for instance it has been shown to release up to 91% of the surface bound 18-MEA with a simple 90 minute room temperature treatment (Meade *et al.* 2008b) Although this high level of delipidation may be advantageous for subsequent surface treatment, drawbacks include the lack of surface specificity resulting in a harsh fibre and/or fabric handle, as well as general fibre damage. Anhydrous *t*-butoxide in *t*-butanol has the advantage of being a highly surface specific treatment, although it removes less of the surface bound 18-MEA. For instance, a two hour treatment at 60°C removes approximately 40% of the 18-MEA.

Hydroxylamine-based delipidation treatments have been reported using a number of different solvent systems (Dauvermann-Gotsche *et al.* 2000). Aqueous hydroxylamine treatment combined with a non-ionic surfactant has been found to remove 70-80% of surface bound 18-MEA without adversely affecting the handle of the treated fabric (Meade *et al.* 2008b). This treatment generates a significant increase in surface wettability and friction, along with low reported oxidation of surface thiols. Addition of surfactant was found to improve the evenness of treatment across a wool fabric surface, without causing any significant change in the amount of 18-MEA and total fatty acids removed by the treatment (Meade *et al.* 2008b). With its moderate, aqueous conditions, this treatment protocol offers a practical route for the application of surface-specific modification to protein fibres.

These delipidation approaches can act as a pre-treatment for subsequent covalent attachment of novel surface modification reagents.

3.6 Covalent surface attachment

Application of surface-specific smart and functional treatments to impart novel properties to protein materials requires an appropriately activated surface with functional groups accessible for modification. The removal of the surface lipid layer of mammalian fibres before the attachment of the new surface, as described earlier in this chapter, enables enhanced accessibility and functionality. Lipid removal, as opposed to surface oxidation alone, has been demonstrated to be critical for the covalent binding of amine-reactive polymer particles (Pille et al. 1998). Covalent surface attachment provides the possibility of high durability to wear and laundering in comparison to conventional technologies based on ionic or other non-covalent forces.

The generation of reactive groups on the surface of protein fibres after disruptive treatments such as UV/ozone treatment or plasma (Xu et al. 2007; Rathinamoorthy et al. 2009) can be used to aid the covalent grafting of surface treatments, such as Ag-loaded SiO_2 nanoparticles (Xu et al. 2007)

Proof-of-principle for the covalent modification of wool fibre surfaces after chemical delipidation with fluorescent and hydrophobic compounds has been demonstrated by Meade et. al. (2008a). Evidence for genuine covalent attachment of a chemical entity to wool surfaces after surface lipid removal was achieved by surface treatment with the fluorescent compound 7-fluorobenz-2-oxa-1,3-diazole-4-sulfonamide (ABD-F). The benzofurazan moiety of ABD-F fluoresces only when covalently bound to a thiol group, and therefore evaluation of the specific fluorescence of bound ABD-F after treatment, performed with fluorescence microscopy, was able to demonstrate successful covalent attachment to surface thiols. The degree of modification was assessed using scanning electron, light and fluorescence microscopy, wettability testing and X-ray photoelectron spectroscopy (XPS), and generally good, even surface modification was observed. These treatments also demonstrated good durability to dye dyeing and laundering durability.

Another class of compounds that have been evaluated for covalent surface attachment is epoxides (Meade et al. 2008a). Epoxides react with a wide range of nucleophiles including amines, thiols and hydroxyls and provide a potential means to exploit all the available nucleophilic groups present in proteins. After delipidation of wool fibre surfaces with hydroxylamine, which significantly increases surface wettability, covalent attachment of fluoroepoxides was shown to restore surface hydrophobicity. The subsequent wetting time, contact angle measurement and XPS analyses were all consistent with the formation of a new covalently bound hydrophobic surface. This novel treatment approach provides a potential route for generating customisable surface hydrophobicity through careful selection of the specific fluoroexpoxide utilised. Once again, these achievements highlight the potential for customising the properties of protein fibre surfaces.

Microparticles and nanoparticles have also been successfully tethered to the wool surface utilising crosslinkers (Meade et al. 2008a). Microencapsulation as a textile treatment technology offers a broad range of applications, but is currently limited for wool and other proteinaceous fibres due to generally poor durability. Covalent attachment of such particles to the fibre surface after delipidation is a potential means to increase treatment durability. To this end, covalent attachment of microparticles and nanoparticles with surface-coated carboxylic and amine surface functionality were investigated utilising both long-range and zero-length crosslinkers. Of the crosslinking technologies evaluated, the crosslinkers 1-ethyl-3-(3-dimethylaminopropyl)carbodiimide, known as EDC, and N-hydroxysuccinimide (NHS), were observed to be the most effective. Particles applied after removal of surface-bound lipids demonstrated increased durability relative to particles applied without prior delipidation.

Once surface thiol groups are exposed via alkali delipidation, maleimide-based treatments also provide an excellent potential route to covalent attachment of new surfaces. Maleimide-derivatives have a high reactive specificity for covalent modification of thiol groups. Thiol-specific gold nanoparticles have been shown to bind (via maleimide reactivity) to thiols exposed on the wool surface following lipid removal by hydroxylamine treatment. This thiol-specific reagent, monomaleimideo nanogold, was used to demonstrate the formation of free thiols on the wool surface and their reactivity towards maleimide-containing reagents via visualisation of the nanoparticles using TEM (Dauvermann-Gotsche *et al.* 2000). The ultimate goal of initial removal of the surface lipids is a customised accessible and exposed surface prior to application of a secondary treatment. Once delipidated in a controlled manner, the exposed reactive thiol surface of the underlying epicuticle and other exposed surface moieties such as hydroxyl and amine groups present targets for permanent attachment of new surfaces. Relative to other forms of surface modification that do not involve prior controlled delipidation, modification *via* attachment to thiol and other nucleophilic functionalities after alkaline treatment appears to be a promising route for providing some significant potential benefits for protein fibres. However, further development of this approach is required before cost-effective, commercially applicable treatments become available.

3.7 Deposits/polymers

Polymers and surface coatings are applied to protein-based fibres for a variety of reasons. The deposition of surface coatings may be examined using XPS or SEM. Polyethylene glycols have been applied to various materials, including wool, to improve thermal storage, resistance to oils, pilling and static charge (Vigo & Bruno 1989). In the pad-cure process, these glycols are crosslinked with dimethylol dihydroxyethylene urea in the presence of an acid catalyst. Super-hydrophobic properties can be obtained for wool and wool blends using *in situ* chemical binding of inexpensive silica and polysiloxane materials, yielding nano-rough surfaces (Zhang & Lamb 2009). To generate improved biomaterials from silk fibroin, cyanuric chloride-activated poly(ethylene glycol) has been applied to give increased hydrophilicity, a smoother morphology (SEM), and increased attachment and proliferation of human fibroblasts (Vepari *et al.* 2010).

Chitosan is frequently applied to protein-based fibres, as it provides additional sites for acid dye adsorption (Ristić *et al.* 2010). Incorporating an enzyme in the alkaline peroxide treatment bath has been reported to enhance wool wettability and the effectiveness of subsequently applied chitosan biopolymer. This also significantly enhanced fibre whiteness. This combination of treatments resulted in a highly shrink resistant machine washable wool fabric. The formation of ionic bonds between the new sulfonic groups generated on the wool fibre surface and chitosan are believed to contribute to this excellent shrink resistance. However, if the enzyme concentration in the peroxide bath is too high, the efficiency of the subsequent chitosan application decreases, as does the shrink resistance (Jovančić *et al.* 2001). Chitosan is also better deposited on the wool surface after plasma treatment (Ristić *et al.* 2010).

Durable polymer coatings in the form of sol-gels may be deposited on fibre surfaces. Inorganic sol gels based on modified oxides of silica, titanium, or other inorganic oxides can form stable layers of small particle size (<50 nm) that improve textile properties on their own merit, and which can be impregnated with customised functional additives such as UV absorbers. These treatments can be applied at low temperature, aside from a brief high temperature curing step (Mahltig *et al.* 2005). Tung & Daoud (2009) reported the coating of wool fibres with inorganic titanium dioxide-based anatase sols prepared using either nitric acid (N-sol) or hydrochloric acid (H-sol). Sol-gel treatments of wool have been reported to result in wool yellowing; Tung and Daoud found that this undesirable effect only resulted after N-sol treatment, likely due to the oxidative nature of nitric acid. In contrast, wool fibres treated with H-sol remained white and exhibited an even surface coating. The UV-absorbing properties of titanium dioxide also resulted in improved UV protection after sol-gel treatment. Most interestingly, treatment with these sols provided a self-cleaning function, with coffee and wine stains disappearing during exposure to UV light. The sol-gel coatings on the wool triggered photocatalytic reactions in the presence of oxygen and water that degraded the chromophores in the food stains (Tung & Daoud 2009). This is a very desirable characteristic in high-value protein-based fibres.

4. Enzymatic modification

A wide range of enzymatic approaches have also been trialled for surface-specific modification of protein fibres. Enzyme treatments offer the prospect of replacing environmentally unacceptable processes with more eco-friendly processes for treating protein fibres. Extensive research and development has been conducted with respect for utilisation of enzymes as antifelting agents for wool, as well as for enhancing the fibre surface colour (Das & Ramaswamy 2006).

Biopolishing, or biofinishing, refers to the application of proteolytic enzymes to the surface of fibres in order to remove protruding fibre components and thereby improve key properties such as pilling, felting and shrink resistance (Durán & Durán 2000). Proteases are the main class of enzyme used for modifying protein fibre surfaces. Proteases are proteolytic enzymes, that is, they act by cleaving peptide bonds and thereby degrading proteins. Utilisation of protease enzymes can improve some physical and mechanical properties of protein fibres such as smoothness, drapeability, dyeing affinity and water absorbency.

As enzymes are sterically bulky at the molecular level, they can often be utilised in a relatively surface specific manner, although general weakening of protein fibres is often observed with such treatments. Treatment with proteolytic or lipolytic enzymes therefore often leads to a perceived softening effect in the fibre, and a reduction in perceived harshness in handle, which can be attributed to a reduction in the fibre bending stiffness through structural protein degradation. A limitation of protease-based treatments is that adsorbed proteases can be difficult to remove from treated fibres, and enzyme retained after rinsing and drying has been shown to cause further degradation under ambient storage conditions (Nolte *et al.* 1996; Durán & Durán 2000).

It has been noted that enzyme concentration and reaction time have a significant impact on the location and level of enzymatic modification at a given pH. Polymers applied for complementary shrink-resist finishing can impede enzymatic action, but, generally speaking, dyeing and oxidative processes leave the fibre more susceptible to enzymatic modification (Nolte *et al.* 1996). In one enzyme trialling experiment, wool yarns were treated with varying concentrations of aqueous protease solutions. Dyeing with madder was then performed on the treated yarns. The observed to decrease in direct correlation to the enzyme concentration used. The wash-fastness of the dye was unchanged by the protease pre-treatments, while the lightfastness was increased (Parvinzadeh 2007).

Enzymatic treatments have been evaluated in tandem with plasma surface treatments. In one study, wool fabrics were treated with low temperature oxygen plasma with and without proteolytic enzymes and examined for their physico-mechanical and dyeing properties. Plasma pre-treatment caused a higher rate of weight loss in the subsequent protease treatment. When wool was dyed with a levelling acid dye, equilibrium dye uptake did not change, but the dyeing rate was observed to increase with plasma pre-treatment followed by protease treatment. With a milling acid dye, the plasma/protease combination treatment was shown to increase both dye uptake and dyeing rate over plasma or enzyme treatments alone. These results appear to indicate that while plasma-induced modification is surface-specific itself, plasma pre-treatment facilitates increased penetration of the enzyme into the fibre. Interestingly, an attempt to polymerise the enzyme with a water-soluble carbodiimide did not observably enhance strength retention (Yoon et al. 1996).

Incorporating an enzyme in alkaline peroxide treatment baths has been reported to enhance wool wettability and the effectiveness of subsequently applied chitosan biopolymer, significantly enhancing fibre whiteness and shrink resistance (Jovančić et al. 2001).

Enzymatic approaches have also been evaluated for improving fibre colour, with discoloration often more severe on and near the fibre surface. Typically this has also involved proteases, which can degrade and remove fibre surface components and thereby increase the overall fibre whiteness (Schumacher et al. 2001). In one study, the efficiency of various enzymes (xylanase, pectinase, savinase, and resinase) in scouring wool was trialled across a range of specialty hair fibres (llama, alpaca, mohair and camel). Significant colour improvement was noted after treatments with xylanase and pectinase (comparable with soap treatment), but not with resinase (Das & Ramaswamy 2006). Colour improvement in terms of resistance to photoyellowing has also been imparted to wool fibres via the laccase-mediated crosslinking of a naturally occurring antioxidant, norhydroguaiaretic acid (NDGA), to the wool surface, along with improved shrink resistance and antioxidant activity (Hossain et al. 2010).

5. Future directions

Global trends indicate that there will be a sustainable and increasing demand for smart and functional textiles. In addition, the move towards natural and sustainable materials continues to gain momentum and is widely expected to be a key driver in consumer decision for decades to come. These factors mean that targeted surface modification of both natural protein fibres and their biomaterial derivatives to provide functional and durable properties will continue to be a growing and exciting area in the global fibre, textile and biomaterial industries.

Recent advances in covalent attachment of new surfaces may provide the platform technologies for a new generation of fibre surface treatments. For protein-based biomaterials, in particular, such durable surface modification provides the potential to overcome current limitations, such as low abrasion and heat resistance. It is anticipated that research performed on the modification of protein fibres, such as wool and silk, will have spill-over application to such biomaterials. However, further research and development is required before these approaches become commercially viable for either natural protein fibres, or fibrous protein biomaterials. In the near future, it is likely that plasma and enzyme-based approaches, with their potential for cost-effectiveness, high throughput processing, and reduced environmental impact, will continue to gain popularity.

6. References

[1] Arai, T., Freddi, Innocenti, R., Kaplan, D.L. & Tsukada, M. (2001), Acylation of silk and wool with acid anhydrides and preparation of water-repellent fibers, *Journal of Applied Polymer Science*, *82*, 2832-2841. doi: 10.1002/app.2137.

[2] Arifoglu, M., Marmer, W.N. & Dudley, R. (1992), Reaction of thiourea with hydrogen peroxide:13C NMR studies of an oxidative/reductive bleaching process, *Textile Research Journal*, *62*(2), 94-100.

[3] Baba, T., Nagasawa, N., Ito, H., Yaida, O. & Miyamoto, T. (2001), Changes in the covalently bound surface lipid layer of damaged wood fibers and their effects on surface properties, *Textile Research Journal*, *71*(4), 308-312.

[4] Bahi, A., Jones, J.T., Carr, C.M., Ulijn, R.V. & Shao, J. (2007), Surface characterization of chemically modified wool, *Textile Research Journal*, *77*(12), 937-945. doi: 10.1177/0040517507083520.

[5] Brack, N., Lamb, R., Pham, D. & Turner, P. (1996), XPS and SIMS investigation of covalently bound lipid on the wool fibre surface, *Surface and Interface Analysis*, *24*(10), 704-710.

[6] Canal, C., Gaboriau, F., Villeger, S., Cvelbar, U. & Ricard, A. (2009), Studies on antibacterial dressings obtained by fluorinated post-discharge plasma, *International Journal of Pharmaceutics*, *367*(1-2), 155-161.

[7] Ceria, A., Rovero, G., Sicardi, S. & Ferrero, F. (2010), Atmospheric continuous cold plasma treatment: Thermal and hydrodynamical diagnostics of a plasma jet pilot unit, *Chemical Engineering and Processing: Process Intensification*, *49*(1), 65-69. doi: 10.1016/j.cep.2009.11.008.

[8] Chen, W., Chen, D. & Wang, X. (2001), Surface modification and bleaching of pigmented wool, *Textile Research Journal*, *71*(5), 441-445. doi: 10.1177/004051750107100512

[9] Das, T. & Ramaswamy, G.N. (2006), Enzyme treatment of wool and specialty hair fibers, *Textile Research Journal*, *76*(2), 126-133. doi: 10.1177/0040517506063387.

[10] Dauvermann-Gotsche, C., Korner, A. & Hocker, H. (1999), Characterization of 18-methyleicosanoic acid-containing proteolipids of wool, *Journal of the Textile Institute*, *90*(3 SI Sp. Iss. SI), 19-29.

[11] Dauvermann-Gotsche, C., Evans, D.J., Corino, G.L. & Korner, A. *Labelling of 18-methyleicosanoic acid cotianing proteolipids of wool with monomaleimido nanogold, Proceedings of the 10th International Wool Textile Research Conference*, Aachen, Germany, 2000, 1-10.

[12] Davarpanah, S., Mahmoodi, N.M., Arami, M., Bahrami, H. & Mazaheri, F. (2009), Environmentally friendly surface modification of silk fiber: Chitosan grafting and dyeing, *Applied Surface Science*, *255*(7), 4171-4176. doi: 10.1016/j.apsusc.2008.11.001.

[13] Demir, A. (2010), Atmospheric plasma advantages for mohair fibers in textile applications, *Fibers and Polymers*, *11*(4), 580-585. doi: 10.1007/s12221-010-0580-2.

[14] Durán, N. & Durán, M. (2000), Enzyme applications in the textile industry, *Review of Progress in Coloration and Related Topics*, *30*(1), 41-44. doi: 10.1111/j.1478-4408.2000.tb03779.x.

[15] Evans, D.J. & Lanczki, M. (1997), Cleavage of integral surface lipids of wool by aminolysis, *Textile Research Journal*, *67*(6), 435-444.

[16] Höcker, H. (2000), Fibre morphology. In Crawshaw, G.H., Ed., *Wool: science and technology*, Cambridge, Woodhead Publishing Limited.

[17] Höcker, H. (2002), Plasma treatment of textile fibers, *Pure and Applied Chemistry*, 74(3), 423-427.

[18] Hossain, K.M.G., González, M.D., Juan, A.R. & Tzanov, T. (2010), Enzyme-mediated coupling of a bi-functional phenolic compound onto wool to enhance its physical, mechanical and functional properties, *Enzyme and Microbial Technology*, 46(3-4), 326-330. doi: 10.1016/j.enzmictec.2009.12.008.

[19] Jocic, D., Jovancic, P., Trajkovic, R. & Seles, G. (1993), Influence of a chlorination treatment on wool dyeing, *Textile Research Journal*, 63(11), 619-626. doi: 10.1177/004051759306301101

[20] Jovančić, P., Jocić, D., Molina, R., Juliá, M.R. & Erra, P. (2001), Shrinkage properties of peroxide-enzyme-biopolymer treated wool, *Textile Research Journal*, 71(11), 948-953. doi: 10.1177/004051750107101103.

[21] Kan, C.W. & Yuen, C.W.M. (2007), Plasma technology in wool, *Textile Progress*, 39(3), 121-187. doi: 10.1080/00405160701628839.

[22] Kutlu, B., Aksit, A. & Mutlu, M. (2010), Surface modification of textiles by glow discharge technique: Part II: Low frequency plasma treatment of wool fabrics with acrylic acid, *Journal of Applied Polymer Science*, 116(3), 1545-1551. doi: 10.1002/app.31286.

[23] Leeder, J.D. & Rippon, J.A. (1983), Modifying the surface of keratin fibres, *Research Disclosure*, 230, 210-211.

[24] Leeder, J.D. & Rippon, J.A. (1985), Changes induced in the properties of wool by specific epicuticle modification, *Journal of the Society of Dyers and Colourists*, 101, 11-16.

[25] Leeder, J.D., Rippon, J.A. & Rivett, D.E. *Modification of the surface properties of wool by treatment with anhydrous alkali*, Proceedings of the 7th International Wool Textile Research Conference, Tokyo, Japan, 1985, 312-320.

[26] Maclaren, J.A. & Milligan, B. (1981), The structure and composition of wool. In, *Wool science - The chemical reactivity of the wool fibre* (pp 1-18), Marrickville, NSW 2204, Australia, Science Press.

[27] Mahltig, B., Haufe, H. & Böttcher, H. (2005), Functionalisation of textiles by inorganic sol-gel coatings, *Journal of Materials Chemistry*, 15, 4385-4398. doi: 10.1039/b505177k.

[28] Meade, S.J., Caldwell, J.P., Hancock, A.J., Coyle, K., Dyer, J.M. & Bryson, W.G. (2008a), Covalent modification of the wool fiber surface: The attachment and durability of model surface treatments, *Textile Research Journal*, 78(12), 1087-1097. doi: 10.1177/0040517507087852.

[29] Meade, S.J., Dyer, J.M., Caldwell, J.P. & Bryson, W.G. (2008b), Covalent modification of the wool fiber surface: Removal of the outer lipid layer, *Textile Research Journal*, 78(11), 943-957. doi: 10.1177/0040517507087859.

[30] Millington, K.R. *Continuous photobleaching of wool*, Byrne, K., Duffield, P.A., Myers, P., Scouller, S. and Swift, J.A., Eds., *Proceedings of the 11th International Wool Research Conference*, Department of Colour & Polymer Chemistry of the University of Leeds, University of Leeds, UK, 2005, 22CCF.

[31] Naebe, M., Cookson, P.G., Rippon, J., Brady, R.P., Wang, X., Brack, N. & van Riessen, G. (2010), Effects of plasma treatment of wool on the uptake of sulfonated dyes with different hydrophobic properties, *Textile Research Journal*, 80(4), 312-324. doi: 10.1177/0040517509338308.

[32] Negri, A.P., Cornell, H.J. & Rivett, D.E. (1991), The nature of covalently bound fatty acids in wool fibres, *Australian Journal of Agricultural Research*, 42(8), 1285-1292. doi: 10.1071/AR9911285

[33] Negri, A.P., Cornell, H.J. & Rivett, D.E. (1993), The modification of the surface diffusion barrier of wool, *Journal of the Society of Dyers and Colourists*, 109(9), 296-301. doi: 10.1111/j.1478-4408.1993.tb01579.x.

[34] Nolte, H., Bishop, D.P. & Höcker (1996), Effects of proteolytic and lipolytic enzymes on untreated and shrink-resist-treated wool, *Journal of the Textile Institute*, 87(1), 212 - 226.

[35] Ottmer, T.C., Baumann, H. & Fuchtenbusch, D. *Physical dyeing parameters of milling dyes with systematically chlorinated and bleached wool, Proceedings of the 7th International wool Textile Research Conference.*, Tokyo, 1985, 131-140.

[36] Parvinzadeh, M. (2007), Effect of proteolytic enzyme on dyeing of wool with madder, *Enzyme and Microbial Technology*, 40(7), 1719-1722. doi: 10.1016/j.enzmictec.2006.10.026.

[37] Pille, L., Church, J.S. & Gilbert, R.G. (1998), Adsorption of amino-functional polymer particles onto keratin fibres, *Journal of Colloid and Interface Science*, 198, 368-377. doi: 10.1006/jcis.1997.5303.

[38] Poll, H.U., Schladitz, U. & Schreiter, S. (2001), Penetration of plasma effects into textile structures, *Surface and Coatings Technology*, 142-144, 489-493.

[39] Prat, R., Koh, Y.J., Babukutty, Y., Kogoma, M., Okazaki, S. & Kodama, M. (2000), Polymer deposition using atmospheric pressure plasma glow (APG) discharge, *Polymer*, 41, 7355-7360. doi: 10.1016/S0032-3861(00)00103-8.

[40] Rakowski, W. (1997), Plasma treatment of wool today. Part 1 - Fibre properties, spinning and shrinkproofing, *Journal of the Society of Dyers and Colourists*, 113, 250- 255. doi: 10.1111/j.1478-4408.1997.tb01909.x.

[41] Ranjbar-Mohammadi, M., Arami, M., Bahrami, H., Mazaheri, F. & Mahmoodi, N.M. (2010), Grafting of chitosan as a biopolymer onto wool fabric using anhydride bridge and its antibacterial property, *Colloids and Surfaces B: Biointerfaces*, 76(2), 397- 403. doi: 10.1016/j.colsurfb.2009.11.014.

[42] Rathinamoorthy, R., Sumothi, M. & Jagadesh, S. (2009), Plasma technology for textile surface enhancement, *Textile Asia*, 40(11), 21-23.

[43] Ristić, N., Jovančić, P., Canal, C. & Jocić, D. (2010), Influence of corona discharge and chitosan surface treatment on dyeing properties of wool, *Journal of Applied Polymer Science*, 117, 2487-2496. doi: 10.1002/app.32127.

[44] Roeper, K., Foehles, J., Peters, D. & Zahn, H. (1984), Morphological composition of the cuticle from chemically treated wool - part II: the role of the cuticle in industrial shrink proofing processes, *Textile Research Journal*, 54(4), 262-270. doi: 10.1177/004051758405400408

[45] Ryu, J., Wakida, T. & Takagishi, T. (1991), Effect of corona discharge on the surface of wool and its application to printing, *Textile Research Journal*, 61(10), 595-601. doi: 10.1177/004051759106101006

[46] Schumacher, K., Heine, E. & Höcker, H. (2001), Extremozymes for improving wool properties, *Journal of Biotechnology*, 89(2-3), 281-288. doi: 10.1016/s0168-1656(01)00314-5.

[47] Shao, J., Hawkyard, C.J. & Carr, C.M. (1997), Investigation into the effect of UV/ozone treatments on the dyeability and printability of wool, *Journal of the Society of Dyers and Colourists, 113*(4), 126-130. doi: 10.1111/j.1478-4408.1997.tb01884.x.

[48] Shao, J., Liu, J. & Carr, C.M. (2001), Investigation into the synergistic effect between UV/ozone exposure and peroxide pad—batch bleaching on the printability of wool, *Coloration Technology, 117*(5), 270-275. doi: 10.1111/j.1478-4408.2001.tb00074.x.

[49] Sugiyama, K., Kiyokawa, K., Matsuoka, H., Itou, A., Hasegawa, K. & Tsutsumi, K. (1998), Generation of non-equilibrium plasma at atmospheric pressure and application for chemical process, *Thin Solid Films, 316*(1-2), 117-122.

[50] Taki, F. (1996), Surface treatments of wool by potassium hydroxide in dehydrated butanol, *Sen'i Gakkaishi, 52*(9), 500-503.

[51] Tung, W.S. & Daoud, W.A. (2009), Photocatalytic self-cleaning keratins: A feasibility study, *Acta Biomaterialia, 5*(1), 50-56. doi: 10.1016/j.actbio.2008.08.009.

[52] Van Rensburg, N.J.J. & Barkhuysen, F.A. (1983), Continuous shrink-resist treatment of wool tops using chlorine gas in a conventional suction-drum backwash, *SAWTRI Technical Report, 539*, 22p.

[53] Vepari, C., Matheson, D., Drummy, L., Naik, R. & Kaplan, D.L. (2010), Surface modification of silk fibroin with poly(ethylene glycol) for antiadhesion and antithrombotic applications, *Journal of Biomedical Materials Research - Part A, 93*(2), 595-606. doi: 10.1002/jbm.a.32565.

[54] Vigo, T.L. & Bruno, J.S. (1989), Improvement of various properties of fibre surfaces containing crosslinked polyethylene glycols, *Journal of Applied Polymer Science, 37*(2), 371-379. doi: 10.1002/app.1989.070370206.

[55] Ward, R.J., Willis, H.A., George, G.A., Guise, G.B., Denning, R.J., Evans, D.J. & Short, R.D. (1993), Surface analysis of wool by X-ray photoelectron spectroscopy and static secondary ion mass spectrometry, *Textile Research Journal, 63*(6), 362-368.

[56] Xin, J.H., Zhu, R.Y., Hua, J.K. & Shen, J. (2002), Surface modification and low temperature dyeing properties of wool treated by UV radiation, *Coloration Technology, 118*(4), 169-173. doi: 10.1111/j.1478-4408.2002.tb00095.x.

[57] Xu, B., Niu, M., Wei, L., Hou, W. & Liu, X. (2007), The structural analysis of biomacromolecule wool fiber with Ag-loading SiO_2 nano-antibacterial agent by UV radiation, *Journal of Photochemistry and Photobiology A: Chemistry, 188*(1), 98-105. doi: 10.1016/j.jphotochem.2006.11.025.

[58] Yilmazer, D. & Kanik, M. (2009), Bleaching of wool with sodium borohydride, *Journal of Engineered Fibers and Fabrics, 4*(3), 45-50.

[59] Yoon, N.S., Lim, Y.J., Tahara, M. & Takagishi, T. (1996), Mechanical and dyeing properties of wool and cotton fabrics treated with low temperature plasma and enzymes, *Textile Research Journal, 66*(5), 329-336. doi: 10.1177/004051759606600507.

[60] Zhang, H. & Lamb, R.N. (2009), Superhydrophobic treatment for textiles via engineering nanotextured silica/polysiloxane hybrid material onto fibres, *Surface Engineering, 25*(1), 21-24. doi: 10.1179/174329408x271390.

Permissions

The contributors of this book come from diverse backgrounds, making this book a truly international effort. This book will bring forth new frontiers with its revolutionizing research information and detailed analysis of the nascent developments around the world.

We would like to thank Prof. E. Perrin Akçakoca Kumbasar, for lending her expertise to make the book truly unique. She has played a crucial role in the development of this book. Without her invaluable contribution this book wouldn't have been possible. She has made vital efforts to compile up to date information on the varied aspects of this subject to make this book a valuable addition to the collection of many professionals and students.

This book was conceptualized with the vision of imparting up-to-date information and advanced data in this field. To ensure the same, a matchless editorial board was set up. Every individual on the board went through rigorous rounds of assessment to prove their worth. After which they invested a large part of their time researching and compiling the most relevant data for our readers. Conferences and sessions were held from time to time between the editorial board and the contributing authors to present the data in the most comprehensible form. The editorial team has worked tirelessly to provide valuable and valid information to help people across the globe.

Every chapter published in this book has been scrutinized by our experts. Their significance has been extensively debated. The topics covered herein carry significant findings which will fuel the growth of the discipline. They may even be implemented as practical applications or may be referred to as a beginning point for another development. Chapters in this book were first published by InTech; hereby published with permission under the Creative Commons Attribution License or equivalent.

The editorial board has been involved in producing this book since its inception. They have spent rigorous hours researching and exploring the diverse topics which have resulted in the successful publishing of this book. They have passed on their knowledge of decades through this book. To expedite this challenging task, the publisher supported the team at every step. A small team of assistant editors was also appointed to further simplify the editing procedure and attain best results for the readers.

Our editorial team has been hand-picked from every corner of the world. Their multi-ethnicity adds dynamic inputs to the discussions which result in innovative outcomes. These outcomes are then further discussed with the researchers and contributors who give their valuable feedback and opinion regarding the same. The feedback is then collaborated with the researches and they are edited in a comprehensive manner to aid the understanding of the subject.

Apart from the editorial board, the designing team has also invested a significant amount of their time in understanding the subject and creating the most relevant covers. They scrutinized every image to scout for the most suitable representation of the subject and create an appropriate cover for the book.

The publishing team has been involved in this book since its early stages. They were actively engaged in every process, be it collecting the data, connecting with the contributors or procuring relevant information. The team has been an ardent support to the editorial, designing and production team. Their endless efforts to recruit the best for this project, has resulted in the accomplishment of this book. They are a veteran in the field of academics and their pool of knowledge is as vast as their experience in printing. Their expertise and guidance has proved useful at every step. Their uncompromising quality standards have made this book an exceptional effort. Their encouragement from time to time has been an inspiration for everyone.

The publisher and the editorial board hope that this book will prove to be a valuable piece of knowledge for researchers, students, practitioners and scholars across the globe.

List of Contributors

Mohamed Abd el-moneim Ramadan, Samar Samy, Marwa abdulhady and Ali Ali Hebeish
Textile Research Division, National Research centre, Dokki, Giza, Egypt

Ashis Kumar Samanta and Adwaita Konar
Department of Jute and Fibre Technology, Institute of Jute Technology, University of Calcutta, India

KenjiOno
RIKEN, Japan
The University of Tokyo, Japan

Daisaku Arita
ISIT (Institute of Systems, Information Technologies and Nanotechnologies), Japan

Yuki Morimoto
RIKEN, Japan
ISIT (Institute of Systems, Information Technologies and Nanotechnologies), Japan

Meritxell Martí, José Luis Parra and Luisa Coderch
Institute of Advanced Chemistry of Catalonia (IQAC-CSIC), Barcelona, Spain

Rattanaphol Mongkholrattanasit
Department of Textile Chemistry Technology, Faculty of Industrial Textile and Fashion Design, Rajamangala University of Technology Phra Nakhon, Bangkok, Thailand

Jiří Kryštůfek, Jakub Wiener and Jarmila Studničková
Department of Textile Chemistry, Faculty of Textile Engineering, Technical University of Liberec, Liberec, Czech Republic

Seung Cheol Yang, Moo Song Kim and Maeng-Sok Kim
Nylon Polyester Polymer R&D Team, Production R&D Center, Hyosung R&D Business Labs, 183, Hogye-dong, Dongan-gu, Anyang-si, Korea

Jolon Dyer
AgResearch Lincoln Research Centre, Christchurch, New Zealand
Lincoln University, Canterbury, New Zealand
Biomolecular Interaction Centre, University of Canterbury, Christchurch, New Zealand
Riddet Institute at Massey University, Palmerston North, New Zealand

Anita Grosvenor
AgResearch Lincoln Research Centre, Christchurch, New Zealand

Printed in the USA
CPSIA information can be obtained
at www.ICGtesting.com
JSHW011328221024
72173JS00003B/96